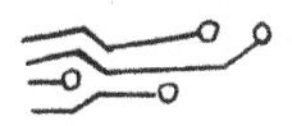

But yield who will to their separation,

My object in living is to unite

My avocation and my vocation

As my two eyes make one in sight.

Only when love and need are one,

And work is play for mortal stakes

Is the deed ever really done

For heaven and the future's sakes.

—ROBERT FROST[1]

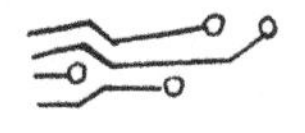

---

1 Last stanza of "Two Tramps in Mud Time" by Robert Frost

INVENTING THE FUTURE

STORIES FROM A TECHNO-OPTIMIST

CORINNA LATHAN, PH.D.

Let's invent the future together!
Cori

INVENTING THE FUTURE
Stories from a Techno-Optimist

ISBN HARDCOVER: 978-1-5445-3536-4
PAPERBACK: 978-1-5445-3535-7
EBOOK: 978-1-5445-3534-0
AUDIOBOOK: 978-1-5445-3537-1

CONTENTS

FOREWORD

BY YO-YO MA, CELLIST

I believe that optimism is a choice. Faced with the hazards and uncertainties of life, humans can be tempted to give in to cynicism or hopelessness. But instead, our story over thousands of years is one of countless intentional acts of optimism—explorations of the unknown, struggles against the odds, unconditional offers of trust, expressions of meaning that we hope will long survive us. Whenever we create—through the arts, the sciences, technology, and so on—we are acting optimistically, in the faith that what we do now may somehow be helpful to the generations after us.

When I first met Cori Lathan at the World Economic Forum's Annual Meeting in Davos, Switzerland, I was impressed by her fundamental optimism—her passion for technology not for its own sake but for its potential to enable people and make positive change. Like many of our most successful creators, she is a "scientist/artist," someone able to see the world both analytically and empathetically. The innovations that she and her collaborators have brought to the world, whether in robotics or neurotech or VR, are grounded in both scientific rigor and a deeply felt compassion for others.

In this book, Cori tells us that "inventions and other artworks" are not only the product of creativity but the source of profound shared experiences. I agree. Both inventions and musical compositions are a source of connection between humans and therefore help us to imagine and build a better world. I hope that Cori's belief in each one of us as a maker and inventor—and her examples of how technology and business can be fundamentally humanist—will be motivating and inspiring to many people, particularly those who are still choosing their path. The world needs more "techno-optimists" and people who believe in the power of culture to make positive change.

As we continue the evergreen work of imagining and building a better future, we will need to be not only great scientists and technologists but also great humanists; Cori Lathan is both. I hope you will take inspiration from the warmth, generosity, and insight of this scientific, social, and cultural entrepreneur.

INTRODUCTION

Technology makes me younger every day. I recently replaced my knee; my vision is better than it's been since the third grade; and I'm working on making my brain better by finally learning a musical instrument—even though my family would probably like me to stop! All in all, I'm optimistic about my future.

But while I was writing this book, I lost both my parents to COVID-19. My mother to the disease, and my father less than a month later to the heartbreak of losing his wife. They had been married for fifty-five years.

I know. That's a lousy way to start a book about inventing the future.

But I get my optimism from my mother. She was the ultimate techno-optimist and early adopter. A mathematician and computer scientist by training, she started teaching robotics in her sixties because to her, programming robots to do cool things inspired her students.

I may have learned optimism from my mother, but I learned to write from my father. He was an English major with a love of opera and acting. Both he and my mom homeschooled my younger siblings through high school and over the years taught at local high schools and colleges.

Together, my parents taught me that math, science, philosophy, and music are essential human pursuits, and they are all needed to create the future we want to see. I think that's why I became an inventor. I wanted to create, but my talents didn't lay in painting or music; they lie in science and mathematics.

I have been fortunate to develop these talents in a career that continues to inspire and challenge me. Yes, I am a neuroscientist, bioengineer, technologist, and entrepreneur, but I also see myself as an artist. Invention has been my medium of choice.

My works of art are the inventions you'll read about in this book—from CosmoBot, the educational robot for kids with disabilities; to 3D Space, the virtual reality experiment in space; to String, the device that stimulates your brain. Like a play or concert, each piece of art is a collaborative effort and a journey of shared experiences.

When I founded AnthroTronix in 1999 along with two students, we thought we were a traditional startup that would make or break it pretty quickly. But more than twenty years later, we have used technology to invent and create for many projects, all with an eye on both the immediate goal and the larger world we want to create.

We started with one simple rule: do cool projects with cool people!

Working with cool people meant finding people who were interesting and ethical. We had a strict no-assholes policy.

Doing cool projects meant using cutting-edge technology to solve problems in the area of human–machine collaboration. In other words, technology should enable human capability.

Let me say that again.

Technology should enable human capability. It should bridge the gap between what we *can* do and what we *could* do.

This book is a collection of those cool projects and cool stories. For instance, the story of CosmoBot, a robot that empowered kids with disabilities to achieve developmental goals, is told in Chapters 1 and 2, which take you through the process of invention and the journey from prototype to product. Chapter 3 tells the story of our instrumented glove, which came from our work with kids and allowed soldiers to communicate with each other and robot team members through digitized hand signals.

Chapter 4 highlights the importance of international cooperation in invention with the story of the 3D Space experiment, where we used virtual reality to study astronauts' perceptual changes during spaceflight. Chapter 5 tells the story of VISUnit, an augmented reality system way ahead of its time, but that, like many inventions, helped bridge current capabilities with future aspirations. Inventions can also be disruptive, changing the way things are done. Chapters 6 and 7 tell stories of healthcare disruption through the DANA Brain Vital mobile medical app and String, a prototype electroceutical to replace pharmaceuticals. As you'll see, finding the right partners is crucial to bringing these ambitious visions to life.

Chapter 8 makes the case for connected cognition and drives home my belief that humans plus technology should always equal more than humans or technology alone. Finally, in Chapter 9, we journey into my lifelong pursuit to recognize and encourage the maker mentality in everyone. The more diversity and inclusion we encourage in STEM fields, the more likely technology will be used to drive toward an equitable future.

In between chapters, we take a break from my story and dive a little deeper into the technology that inspired our inventions. Just like inventions, these short reflections bridge the past and future of technology.

Embedded in this book is, of course, advice for entrepreneurs, tips and tricks for inventors, and lessons learned from a woman in STEM. But mostly, I'm writing for the creatives out there. I want to share how I got here, but your journey will be as unique as mine. Either start at the beginning, or cut straight to a project that speaks to your passion.

Are you a maker, creator, inventor, artist? Are you curious about technology? Do you want to make the world a better place?

Then this book is for you. From my parents and from me.

IF you think about solutions differently,
THEN you can invent the future.

CHAPTER 1

Robots for Kids

THE STORY OF COSMOBOT

JOEY IS ON the autism spectrum. For many months now, his therapist has been trying to get him to practice a form of social behavior called joint attention. For example, if I point to a box and say, "Look at that," you would likely follow the line of sight from my finger, realize what I am referring to, turn your head to look, and maybe even respond, "Yes, I see it." Joint attention enables us to focus *together* on things and people, so it is key to communication and learning. But kids on the spectrum may find it challenging to pay attention to a person and an object at the same time, which makes learning joint attention difficult.

So my team and I are here to introduce Joey to our robot, CosmoBot. CosmoBot is about eighteen inches high and can move its arms and roll around but has very limited facial expressions. We place CosmoBot in front of Joey. CosmoBot points to a box and says, "Hey, look at that." Joey ignores it. In contrast, Joey's sister, sitting beside him, immediately looks at the box—joint attention comes easily to her.

We try again. "Hey, look at that." No response from Joey.

We try again. And again.

After about three minutes, which equals a *lot* of tries, CosmoBot points at the box and again says, "Hey, look at that." This time, Joey looks at CosmoBot. Progress! He's not looking at the box yet, but now we have his attention.

A few tries later, when CosmoBot points to the box and says, "Hey, look at that," Joey reaches out and takes the robot's hand. Getting closer.

We keep trying. CosmoBot points and says, "Hey, look at that." Joey smiles.

Five minutes and counting.

CosmoBot points to the box and says, "Hey, look at that." This time, Joey looks at the box. Both Joey and the robot are looking at the box *together*—the very picture of joint attention! As a reward, the box opens and a robot toy walks out, to Joey's delight.

In this moment, all the research and what-ifs and trial and error pay off. Joey's connection with CosmoBot may enable him to relate to others more easily, transforming his ability to learn and communicate. This is exactly what I've dreamed of—inventing technology that expands human ability. But how did that lead me to CosmoBot and Joey? Let me back up...

The Perfect Job

When looking for my first job at the end of graduate school, I was faced with several choices. One was to continue my research in neuroscience and do a postdoctoral fellowship. A very well-known laboratory in Portland, Oregon, had offered me a position that would begin a natural career path. Portland was also a great city in a part of the country I was excited to explore. This was appealing, but I wasn't sure I wanted to continue the same line of work. I was ready for a change.

Another option was to follow an engineering path and take an industry job in the Midwest at an exciting new company working on robotic exoskeletons. This would put me at the cutting edge of human-technology interaction. Also appealing, but I wasn't sure if the industry was the right fit for me.

A third job possibility seemed just right: a faculty position in the up-and-coming field of biomedical engineering at the Catholic University of America. Because biomedical engineering was such a new field, none of the professors had a degree in it, so my lack of one wasn't an issue. In fact, my eclectic background in math, psychology, neuroscience, and aerospace engineering fit right in. I had just spent ten years in academia and felt comfortable in that environment. I also liked that CUA was a smaller university with a balance of research and teaching. In addition, it was near my family, who were based in western Maryland—another plus.

And so, in August 1995, I started my first job: Assistant Professor of Biomedical Engineering at the Catholic University of America.

How Can Technology Help?

I had learned about the CUA position in a fortuitous way, after reading about a project that used spacesuit technology to help patients living with multiple sclerosis (MS). The suit used a network of tubes to distribute water to heat or cool the suit. That system was adapted to try and improve circulation in the MS patients. I was intrigued by the application of aerospace technology to a medical problem and wanted to know more. It turned out that one of the project sites, the National Rehabilitation Hospital (NRH), worked closely with CUA. Spacesuit cooling technology for MS patients never made it into medical use, as far as I know, but the concept of technology transfer, which grabbed my interest when I read the article, underlies many innovations.

Technology transfer simply means that technology developed for one purpose can be used for an unrelated purpose. For example, imaging technologies developed for Mars exploration are what enable the panoramic photos we now take for granted in real estate walk-throughs. Research in artificial robotic muscle systems have improved the sensitivity and functionality of human prosthetics.[2]

A related concept, dual-use technology, particularly applies to new and emerging technologies with both military and civilian applications. For example, at the time, robotics weren't in widespread use anywhere, so almost any progress was applicable in multiple settings. My new research lab at CUA, called the Computer-Human Assist-Oriented Systems (CHAOS) Lab, was funded by grants to support dual-use technology and was part of a government-funded research center at NRH focused on the emerging technologies of robotics, sensors, and virtual reality. My first task was to choose which of the center's three main populations the CHAOS Lab would serve: (1) older adults (e.g., many of the patients in stroke rehabilitation), (2) the workforce (e.g., persons with disabilities needing job accommodations), or (3) children (e.g., those living with cerebral palsy and other challenges).

As an engineer, my choice was clear. Engineers specify functional goals and then utilize technology to achieve them. Inventors do this in a way that no one has done before. So for my purposes, children were an engineer's dream!

Developmentally speaking, all children have essentially the same design specs and critical milestones. They may reach those milestones at different rates and in different ways, but all need to learn

---

2 Publications and Graphics Department, NASA Center for AeroSpace Information (CASI), *NASA Spinoff: 50 Years of NASA-Derived Technologies (1958–2008)* (Washington, DC: NASA, 2008), https://spinoff.nasa.gov/Spinoff2008/pdf/spinoff2008.pdf.

things like cause and effect, receptive and expressive language skills, and social cognitive skills. Furthermore, we expect children to gain more function and capability as they grow, and the right technology can facilitate that.

In contrast, although the other two choices were compelling, they were less exciting to me at the time. The elderly tend to lose function or have inconsistent function over time, so designing technology for them would be much harder. Viewed purely through an engineer's eyes, their design specifications are not consistent or predictable. Creating solutions for workforce accommodations would offer many interesting design challenges, but accommodations tend to be very targeted to an individual and their job, and I was curious about more scalable solutions.

Of course, other professors at the center were still going to work with those populations. But I got the kids.

Rock 'n' Roll Science

My team at the CHAOS Lab started working with kids who were referred to us through the NRH. In our first year, our idea was to work with as many children with disabilities as we could, as design partners, and create lots of cool prototypes. These kids had a variety of developmental goals, aided by exacting and strenuous efforts from physical and occupational therapists.

One of the hardest parts of a therapist's job is motivating a client to practice the prescribed therapy. I was convinced we could make therapy more fun and motivating with technology. We sometimes called our approach rock 'n' roll science because of the enthusiasm, imaginativeness, and creativity that drove our efforts—often to dramatic results!

Meanwhile, because this was still academia, we were expected to present and share our work. But where? At the time, tech conferences and medical conferences didn't overlap much. Then we found an amazing new conference, Medicine Meets Virtual Reality (MMVR), which brought technology innovators together with medical professionals.

At MMVR, I met kindred spirit Dave Warner, an MD/PhD and co-director of the Center for Really Neat Research (CRNR) at Syracuse University. It was the start of a great partnership that exists to this day.

Working with the CRNR, we were able to take our work at the CHAOS Lab up a notch. Dr. Dave's team had developed tools that easily enabled analog to digital conversion, which immediately expanded the possibilities for the CHAOS prototypes. Analog to digital conversion meant that we could take any sensor, attach it to a standard radio plug, and plug it into their custom boxes, called THG1 and THG2 (yes, named after the Dr. Seuss characters!). Those boxes plugged directly into a computer's serial port (the precursor to today's USB and micro-USB ports). We could then map each sensor to a function on a keyboard or to an external device (e.g., a mouse) that was also connected to the computer.

To understand how powerful this technology was, let me tell you about Chris,[3] a fifteen-year-old participant in our lab who had cerebral palsy (CP). CP is a group of neurological disorders that can impair your motor system and sometimes your cognition as well. Chris was affected in both ways, so we had to enable them to physically interact with the computer as well as meet their level of understanding. Specifically, we needed to learn what on the computer screen would motivate them to react and give them the right tools to express a reaction.

So we embedded inexpensive light-sensitive sensors in small Lego building bricks. We could then place the bricks anywhere on

---

3 I've changed the name of Chris and other young participants to protect their privacy.

the table within Chris's reach. Chris didn't need fine motor control to activate the sensors—they didn't even need to touch the bricks. We set the sensor threshold so that as soon as they moved a hand near a brick, the light would activate. We then mapped each sensor to a keyboard command that controlled an action on the computer screen.

Chris was a baseball fan, so we created a game with the actions. If they moved their hand near one brick, a player on-screen swung a bat. If they moved their hand near a different brick, a player ran to first base. Other bricks enabled them to have a player to run to another base or stay put. By interacting with the bricks, Chris was able to make decisions in a game that they were familiar with and loved. Very quickly, Chris was meeting physical therapy goals *and* enjoying their favorite game!

Another participant in the CHAOS Lab was Jaylin, a young child with mobility issues. Jaylin's therapist was trying to increase Jaylin's range of shoulder motion, so we designed sensor-activated games to encourage them. We used tilt sensors, meaning the sensor triggered a computer action when moved from a horizontal to a vertical position, for example. For one game, we put the sensors on Jaylin's arm, so every time they raised their arm, an onscreen character would jump in the air—a simple and satisfying result! For another game, we connected the sensors to a remote-control car. Jaylin practiced raising and lowering their arm to control the car's movement.

Using sensors and some imagination, we were able to motivate kids to reach their goals and have fun doing it. Their enthusiasm spoke volumes, and we were also getting great feedback from the therapists. For example, Jaylin's therapist no longer had to focus on just getting Jaylin to raise their arm, saying something along the lines of, "Jaylin is so happy to raise their arm now to play the game. I don't even have to ask them!" The therapist could now spend more time monitoring and modifying the therapy itself. Onward!

Fun Isn't Enough

The parents of our kid participants at the CHAOS Lab had a different perspective. One parent's response: "I love that they are having fun doing their physical therapy. But because of their disability, I have to make the choice every day between spending time on homework *or* physical therapy." Other parents echoed this statement in many different ways, such as one parent who said, "Many times, homework takes priority and I can't get to their therapy. Can't we do both at the same time?"

BOOM! These comments hit me like a lightning bolt. I wasn't solving the real problem.

It was clear that we needed to move from rock 'n' roll science to a more holistic approach. The team did some brainstorming and came up with the idea of building our own platform that could be programmed to motivate the child to learn while doing their physical therapy. We adopted a key functional requirement: our solution had to facilitate a child's developmental goals by engaging them both cognitively *and* physically. And what better way to engage a child in both learning and movement than a robot?

First Prototype: JesterBot

While we were learning from participants at the CHAOS Lab, others were doing cutting-edge research that helped sharpen our focus. A wealth of research by Dr. Cynthia Breazeal's team at the MIT Media Lab showed that a 3D animatron was more engaging than a computer-based character.[4] Around the same time, Dr. Allison

4 C. Kidd and C. Breazeal, "Effect of a Robot on User Perceptions," 2004 IEEE/RSJ International Conference on Intelligent Robots and Systems (IROS) (IEEE Cat. No.04CH37566), Volume 4 (2004), 3559–3564.

Druin, director of the Human-Computer Interaction Laboratory at the University of Maryland, was developing storytelling robots along with her team of kid inventors.[5] Their efforts reinforced my hypothesis that robots could be a powerful cognitive and motor developmental tool for kids.

Our first iteration was bare bones. A toy remote-control car, similar to the one Jaylin used in the CHAOS Lab, provided the wheeled base, or chassis. Allison's team of kid inventors made us a foam character we called JesterBot, both because of its comical jester-like appearance and because it was controlled through gestures. JesterBot's foam body fit over the car chassis, and JesterBot's head concealed two additional motors that controlled its head and arms.

To activate the motors, we used remote-control frequencies (Bluetooth was not yet widely available) and rewired the controllers so that the tilt switches (the same kind we'd used with Jaylin) could activate the robot's motors. We were ready to put JesterBot to work.

Proof of Concept

We started working with young children with CP who struggled with the developmental concept of cause and effect—the idea that your actions can cause a response in the world around you. For example, when a baby shakes a rattle, they hear a noise and realize quickly that they can produce a noise again. And again. But if you don't have the motor control to shake the rattle, it's harder to internalize the concept that your actions can cause reactions.

One little boy we worked with, Mikey, needed to learn cause and effect. We wrapped a sensor around Mikey's arm and asked him to

---

5 Allison Druin, James A. Hendler, and James Hendler, eds. *Robots for kids: exploring new technologies for learning*. Morgan Kaufmann, 2000.

Mikey learns cause and effect with JesterBot. (Photo courtesy of AnthroTronix.)

raise it. The sensor was mapped to JesterBot's arm so that if Mikey raised his arm, JesterBot would raise its arm too.

At first, this approach didn't get us very far. If I raised my arm, Mikey was able to raise his arm, but he didn't seem to connect his behavior to JesterBot's action. Mikey was simply mimicking me, much like a baby who sticks out their tongue at you or gives you a high five if you hold up your hand first. Mikey didn't understand that when he raised his arm and then the robot arm raised its arm, he had made that happen. We needed to make it clearer that Mikey had power in this world!

Next, we mapped the sensor on Mikey's arm to *all* the motors on the robot. This time, when I raised my arm and Mikey raised his to mimic me, the robot danced! Mikey's eyes lit up and he raised his arm again. And again. Completely on his own. His parents, the therapist, and my team were astonished! This was rock 'n' roll science at its best but this time with a therapeutic robot instead of a game.

We also worked with five-year-old Jessie. Jessie had developmental delays that affected her social and cognitive development. The therapist wanted to use JesterBot to help Jessie build expressive language skills. In other words, the therapist wanted to get Jessie to talk.

Jessie sat with her mother and JesterBot in one room while the therapist spoke from another room, using JesterBot as an avatar. Jessie's mother showed her pictures of objects, and the therapist, through JesterBot, asked questions to encourage Jessie to respond.

Although this approach was somewhat successful in eliciting responses, a much bigger breakthrough happened. Jessie built a relationship with JesterBot, touching and hugging the robot. Then she went home and did the same thing, *for the first time*, with her stuffed animals and dolls.

Somehow, interacting with JesterBot had unlocked the development goals associated with social-cognitive interaction. Typically, children reach this developmental goal around two or three. At five, Jessie just took a little longer.

The Next Iteration: CosmoBot

By this time, thanks to input from our kid inventors, we had developed the next version of our robot, named CosmoBot. Whereas JesterBot appealed to very young children, we designed CosmoBot to appeal to older kids as well. Instead of a simple foam body, it had articulated arms and legs and a friendly face, and it included an embedded device called an iPAQ, made by Compaq. This was a handheld computing device that predated the iPhone by ten years and cost about $500.

The Motion Analysis Laboratory at the Mayo Clinic offered us a great opportunity to test the feasibility of CosmoBot on a broader

The evolution of JesterBot to CosmoBot. (Photo courtesy of AnthroTronix.)

scale than was possible at the CHAOS Lab. The Mayo Clinic is one of the top clinical and research organizations in the United States, and we were fortunate to find an innovation partner in Dr. Krista Coleman, a research therapist in the Motion Analysis Lab. Dr. Coleman recognized the potential of CosmoBot as a therapeutic tool, with sensors capable of detecting movement smaller than is visible with the human eye and providing rapid and engaging feedback. She hypothesized that children performing goal-directed tasks receiving feedback from CosmoBot would have better outcomes than performing the same goal-directed tasks following traditional therapy. As she put it, "The mind and body are connected, and if you supply the mind with feedback from correct attempts to produce body movement, you will engage them both and the whole will be more than the sum of its parts."

To test this idea, children with physical limitations due to neurological involvement (such as those with cerebral palsy) participated in a six-week period of twice-weekly traditional therapist-guided intervention and a separate six-week period of twice-weekly intervention using CosmoBot. Each child had a specific movement goal, such as extending their wrist to help in opening their hand. Therapists are great at making this fun and motivating for a child during traditional therapy. However, CosmoBot can provide faster feedback of a correct movement attempt and can do so in a more goal-directed manner. The primary motivator was to make CosmoBot move enough to push a ball into a goal.

At the beginning and after each intervention period, each child was evaluated. Children were able to increase their wrist movement following both interventions. But the amazing result was that following therapy using CosmoBot, the children also increased movement *beyond* the wrist. For example, their hand fine motor control and their ability to coordinate movements improved as well. These results were published by Dr. Coleman and really changed the way therapists viewed possibilities for technology like CosmoBot.[6] They were able to see the benefits of a more holistic strategy—motivating movement rather than just isolating one muscle or joint. This was particularly important for kids in a tremendous state of neural plasticity, or brain development.

But Is There a Market?

We were experiencing success with JesterBot and CosmoBot, and parents and therapists wanted it! This raised an important question to us as inventors. Is there a market? In other words, are there enough customers who will buy our invention at a price we can offer? And who are those customers? The therapists or the parents? Well, it depends on who pays. We had been so excited about working with kids that we hadn't thought about these practicalities.

One thing to consider was price point. At the time, the market for assistive technology—meaning technology that assists persons

---

6 Krista A. Coleman Wood, Corinna E. Lathan, and Kenton R. Kaufman, "Development of an Interactive Upper Extremity Gestural Robotic Feedback System: From Bench to Reality," *2009 Annual International Conference of the IEEE Engineering in Medicine and Biology Society*, 5973–5976, doi:10.1109/IEMBS.2009.5333523.

with disabilities in achieving activities of daily living such as dressing, cooking, or accessing a computer—was pretty limited. For instance, a standard piece of technology used by many kids we worked with was called a BIGmack. It was a large button that could be mapped to a few simple actions. Many kids used the BIGmack as a computer mouse, with one switch to move the cursor and another to click. This type of device cost around $50 and required almost no training to use.

The good news was, CosmoBot could provide way more functionality. The bad news was, potential customers were comfortable with a low-cost, low-tech solution. It is hard to get people to buy a product when they already have a cheaper, more recognizable solution.

This was especially true because at the time, robotic technology was not even on most people's radar. A few simple robotic toys had hit the market, like Hasbro's My Real Baby doll and Sony's robotic dog, Aibo, but there was nothing yet in the assistive technology world.

Our best guess was that at scale, we would be able to sell CosmoBot for $499, not including the iPAQ. We thought this price point would appeal to clinics, because $500 was a common threshold for equipment purchases. In many cases, individual therapists could purchase an item and be reimbursed for anything up to that amount. Though it would clearly be an investment, some families could buy it directly at that price as well. In addition, the Americans with Disabilities Act required educational institutions to provide assistive technology for kids with disabilities so they could have the same access to education as typically developing kids. Purchasing a CosmoBot would enable them to meet that requirement.

In short, our initial market research suggested that there was a need and a desire for a learning tool like CosmoBot, so we decided to try and bring it to market. However, the journey to commercialize an invention turned out to be much less straightforward than we anticipated and ultimately led to an even more ambitious product, Cosmo's Learning System. In the next chapter, you'll find out how.

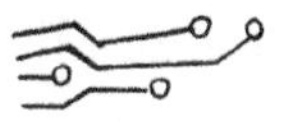

INSPIRATION AND REFLECTION #1
Social Robots

Social robots, like CosmoBot, build relationships by communicating with us rather than just serving a function. Many social robots spend their entire "lives" in academic research labs, and they don't get as much recognition as I think they deserve. So I want to tell you about some of the leaders in this field who inspired our work and the amazing creations they pioneered.

One rock star roboticist is Professor Maja Matarić, head of the Interaction Lab at the University of Southern California. The Interaction Lab was the first to deploy real-world studies with socially assistive robots, a term Dr. Matarić and her team coined. For example, her research team has helped children with autism improve their imitation skills by working with a humanlike physical robot called Bandit and improve their social skills with a tabletop robot called Kiwi. They also used social robots with infants to encourage movement through observation and imitation. Another study saw older adults improve their rehabilitation goals with Bandit's help. Interestingly, the researchers found that people liked Bandit even more when it explained, "Sometimes my mouth just doesn't work. I'm sorry about that."

Another visionary is Dr. Justine Cassell, a professor at Carnegie Mellon University and Director of the ArticuLab, who developed SARA—a Socially Aware Robot Assistant. SARA personalizes its interactions with individuals and collaborates with its human users. SARA is an AI-powered virtual social robot. At the World Economic Forum in Davos, Switzerland one year, attendees could ask SARA which sessions they should go to based on their interests. SARA would also suggest people they might want to meet. Dr. Cassell has the best definition of AI that I've heard. She says, "AI is not a technology; it's a technique for understanding people and making machines act the way people do." She goes on to say that this allows us to make machines that know how to collaborate with people rather than replacing them. So using Dr. Cassell's definition, the role of machine learning and artificial intelligence in social robotics is clear: to close the loop on building the relationship between human and machine.

I've mentioned Professor Cynthia Breazeal, whose research at the MIT Media Lab inspired our first robot prototype, JesterBot. Dr. Breazeal also brought to market one of my favorite robots, a personal assistant called Jibo. Her vision for Jibo was ambitious. It acted as a wingman as you cooked, hosted a party, and set up a dinner date. If the promo video[7] is still on the internet, it's well worth a watch. Jibo came on the market in 2017 and was gone by 2019, as it didn't quite live up to its billing as a personal assistant. However, the small robot gained a significant following through its endearing social interactions. My kids loved asking Jibo to tell a joke or to twerk! This lamented star may yet have a resurgence since as of this writing, Jibo.com is back online.

Given the continuing advances in the market, I think the future for social robots is bright. A sophisticated robot friend that you love even more for its imperfections may be closer than you think.

---

7 "JIBO: The World's First Social Robot for the Home," YouTube video, 3:13, September 30, 2015, https://www.youtube.com/watch?v=H0h20jRA5M0.

IF you give kids the right tools,
THEN you enable them to reach their full potential.

CHAPTER 2

Software for Education

THE STORY OF COSMO'S LEARNING SYSTEM

I GET OFF THE plane in Hong Kong and meet two of my colleagues in the airport. We board a train to Shenzhen, China. We've chosen a small factory there to do a run of 500 devices for our new product, Cosmo's Learning System (CLS). These devices will serve as an alternative to a computer keyboard and are critical to our success. Kids will be banging on them and yelling into them to control Cosmo, the star character in CLS, so the devices need to be both easy to use and durable.

Our train passes hundreds of factories and probably thousands of small distribution shops before we reach our destination. In some forgettable hotel room, I toss and turn all night, trying to calm my anxiety and prepare myself for everything we need to accomplish.

I've risked a lot already and am completely out of my depth, having never manufactured a product before.

The next day, I realize I could have slept soundly. The factory is bright and well organized, the equipment is state of the art, and the employees work together with ease and flawless precision. But how did I end up on a factory floor examining workflows? Let me back up...

From Academia to Incubator

It turns out that academia was not a great fit for me. I liked working with students but not in a traditional college course format. My first semester, the department chair asked me to teach an 8:00 a.m. class on differential equations. Not it! I didn't want to get up that early for a boring math class and I didn't think the students would either. It's not that I think differential equations are boring in general, but I disagree with an entire class being dedicated to the topic when it could be taught in the context of actual engineering! To inspire engineering students, I wanted to teach an applied course, such as "Innovations in Biomedical Engineering," and include differential equations as a tool. But instead, they switched the math class to 1:30 p.m. and I reluctantly agreed to teach it. But I hated it.

What I loved about teaching was the more advanced courses in rehabilitation engineering and human–computer interaction. These were maker courses, where we could brainstorm and implement solutions for real-world problems. At that time, the term "maker" hadn't yet come into vogue, but "user-centered design" was taking off. Inventing and using technology in creative ways to solve real problems with the user at the center was what excited me and the students, not learning mathematical tools for the sake of a graduation requirement.

There was also this absurd thing called tenure that my life was supposed to revolve around achieving. Tenure essentially grants a professor lifetime employment at the institution, regardless of the quality of their research and teaching. The designation was originally put in place to ensure academic freedom of faculty members, but in some cases it means that tenured faculty hangs around as dead weight.

The main metric for tenure seemed to be how many papers a person wrote, though the actual number required was never documented. Tenured professors were also generally the only ones trusted to run a department. But the department chairs and deans had vastly uneven levels of ability to manage a bunch of egocentric faculty members.

I was uninspired, to put it mildly, by the leadership.

I was also the only female professor in the entire school of engineering. Granted, it was a small school, with about thirty or so professors covering four departments including mine, biomedical engineering. But as the only woman, I was asked to serve on almost every committee, and since I hadn't yet learned the delicate art of saying no, I burned out pretty quickly.

AnthroTronix Is Born

In 1999, I received a summer faculty fellowship from NASA to work at the Space Systems Laboratory at the University of Maryland (UMd). One day, while I was visiting some collaborators there, I saw a building going up across the street. I learned that it was a business incubator—at the time, a very new concept. University-based incubators give entrepreneurs the infrastructure needed to get a company off the ground by providing offices, administrative support, and mentoring.

From the moment I learned what a business incubator was, I wanted to apply. Despite the fact that I now had an established research lab at CUA and appointments at UMd, Georgetown University School of Medicine, and the NRH, I felt the need to build something new. I knew I could get closer to my vision for assistive technology in an entrepreneurial setting. After four years as a professor, I was ready for something different.

I started the company with two students, Mike Tracey and Jack Vice. Mike was a doctoral student at CUA in biomedical engineering and Jack was a former marine, finishing his undergraduate degree in physics at UMd. We needed a company name that captured our vision of human-centered technology innovation, and our colleague from Syracuse, Dave Warner, suggested combining "anthro," meaning human, and "tronix," meaning instrumentation or technology.

In July 1999, AnthroTronix was born and I took a leave of absence from Catholic University. We were accepted into the UMd business incubator, and their program gave us four years to get our company off the ground.

With that news, I told the leadership at CUA that I would only come back if the biomedical engineering chair died or left and if the university voted early on my tenure. The chair did *not* die (fortunately), but he did leave, and they voted to give me tenure. Because they'd met my "demands," I felt I had to go back. But after just a month, I was chafing at the bit. I wanted to use technology to solve problems for real people, not research how to do it. For better or worse, I permanently left academia and set about becoming a CEO.

Making a Business Case

As you learned in Chapter 1, we knew CosmoBot was a great product: it had performed well in trials, and our new company was ready to try and sell it commercially. But to do that, we needed money to get AnthroTronix off the ground.

We put together a document called a private placement or a nonpublic offering, which allowed us to raise money from individuals. This approach is often called raising money from "angels"—and sometimes, "friends, family, and fools"! But we couldn't just take money from anybody. Reputable private placements are governed by laws that protect the "fools" by requiring certain qualifications. Generally, the investor must meet a certain threshold in terms of assets or net worth. Given that most startups fail, these requirements are important. They ensure that risky investments don't take advantage of people who may not be able to weather the loss.

Despite the clear need for CosmoBot and even our sound business model, there were many obstacles. One was the lack of maturity of the technology. Another was the difficulty of manufacturing. Lack of maturity is part and parcel of an innovative product, but it also means the technology is still experimental. People don't trust it yet and may be reluctant to use it. Difficulty in manufacturing comes into play when you have a complicated new design that no one has made before. There is a lot of potential for the technology to fail if it's not made properly.

A third obstacle, related to the first two, was cost. We highly underestimated the amount of money needed to design and manufacture even a minimally viable product. I'm reminded of the scene from the movie *Austin Powers*, where Doctor Evil wakes up after years of cryogenic sleep and proposes holding the world ransom "for one *million* dollars." After his henchmen catch him up on the modern economy, he adjusts. "Okay, we will hold the world ransom for one *hundred billion* dollars." That's about how far off we were! We were trying to raise $1 million in our initial round. We didn't need $100 billion, but given the immature technology and manufacturing challenges, $10 million might have been a more realistic target to start.

But that's hindsight, and we were about to face a couple more obstacles: the investors themselves and forces beyond our control.

Meeting with Investors

It was September 10, 2001. I had flown back the night before from a consulting job in Europe because we had our first investor pitch that evening to a well-known angel group in the DC area called The Dinner Club.

We were optimistic as we had been getting great press for our work in *Forbes* and *Time* magazine. Everyone loves a news story about robots and kids! We had also raised our first $100,000 from friends and family (no fools).

While the investors dined in the private room of a well-known DC restaurant, my Chief Financial Officer, Carl, sat with me in the hallway outside, waiting. After at least an hour, we were called in. Our audience was about a dozen dinner club members—all men—basking in the enjoyment of a good meal and yet another round of drinks. They loved our short presentation and thought JesterBot was adorable. But no one responded to my assurances that "all kids have the same design specifications." My suggestion that the Americans with Disabilities Act ensured funding for our product seemed to fall on deaf (and drunk?) ears.

One man commented, "It's so sweet you're working with kids." Another drawled, "You must have a big heart." Gag.

Even my CFO, Carl, an older white male with a tremendous amount of Silicon Valley experience including selling two companies, had no luck in redirecting their attention to our business plan. The group spent a few minutes playing with our robot and then dismissed us.

We paced in the hallway, hoping they were deliberating the pros and cons of our proposal. After another hour, when they'd finished their dessert and drinks, they filed out. Some gave us polite nods and said again how "wonderful" it was what we were doing—as if we were dabbling in charity work. The fund manager said he would get back to us in about a week.

Carl said he had never experienced such an appalling lack of respect for his time or credibility. Unfortunately, like many female entrepreneurs, I became all too familiar with it. Raising money is notoriously difficult for female founders even today.

Maybe we shouldn't have expected to beat the odds, but that all became moot the next morning.

On September 11, 2001, after the terrorist attack that took many lives, the investment community shut down and didn't recover for a decade. In the year prior to 9/11, venture capital firms (VCs) raised $112 billion from investors for new and follow-on funds. In the years following, VCs never came close to that kind of fundraising activity. In 2010, for instance, US VCs raised only about $13 billion.[8] Many small businesses folded after that catastrophic event and many more never got off the ground.

Still a Hard Sell

After the investment funding fallout from 9/11, AnthroTronix was about to go under after only two years in business. But earlier in 2001, we had learned about another opportunity, the Small Business Innovation Research (SBIR) program. All government agencies that fund research, such as the National Institutes of Health (NIH), the National Science Foundation (NSF), and even the Department of Defense (DoD), have to give a small percentage of those research funds to small businesses. We applied and made the case for CosmoBot as an education and therapy platform for kids with disabilities.

These grants have a long lead time and are not easy to obtain, as they involve a lot of paperwork, and must be written almost like scientific papers—not something most entrepreneurs are skilled at or interested in doing. But coming from academia, grant writing was

8 Tom Stein, "A Decade Later: VCs Look Back on Impact of 9/11," *Venture Capital Journal*, September 9, 2011, https://www.reuters.com/article/us-venturecapital-911-vcj/a-decade-later-vcs-look-back-on-impact-of-9-11-idUSTRE7883W920110909.

second nature to me. So in November 2001, AnthroTronix received our first SBIR grant from the NSF. Shortly thereafter, we received a similar grant from the NIH. We were back in business!

The first part of an SBIR grant is determining the feasibility of a product idea. With the results from the Mayo Clinic study as well as other related studies, we had shown both the technical and clinical feasibility of using CosmoBot with kids with disabilities. Rather than positioning it as a consumer product, we focused on how it could function within medical programs and research facilities to help kids with disabilities practice therapy while also working on developmental goals. In other words, working on their mind and body at the same time.

But the second part of the SBIR funding evaluated the commercial feasibility of our product. As we had learned when we sought private funding, this was going to be the hard part. We had to find the intersection of customer need, technology maturity, and market readiness.

Customer need, at least, seemed clear and well segmented. Educational technology could benefit all kids, but we were specifically targeting the millions who could benefit from and be supported by the Americans with Disabilities Act.

Technology maturity and market readiness, however, were still weak points. Educational or therapeutic robots were a long way from being affordable, and potential customers did not yet see them as indispensable. So what *was* indispensable to our target market of kids with disabilities, and what was our "value-add"?

The Pivot: Cosmo's Learning System

The desire to impact access to education through assistive technology was by far the biggest driving force behind our efforts. We believed technology could enable kids to reach their full potential if we could deliver it at critical developmental periods in their lives. However, we were struggling with bringing our robot to market.

To succeed in meeting our goal, we had to pivot. That's a term you hear a lot from entrepreneurs when we realize that something has changed with our product, customers, or market and we need to change strategy. With disruptive technology, pivoting is almost inevitable. Disruptive technology is technology that fundamentally changes an aspect of how people behave. With CosmoBot, we were disrupting the assistive technology field by combining educational access and therapy. So we pivoted our product ideation from a stand-alone robot to an educational software platform: Cosmo's Learning System.

Cosmo's Learning System (CLS) combined learning games and activities to help with a host of developmental goals. To evolve from robot to learning system, we needed to design a simple, sturdy interface that enabled users to interact with the computer. We called it Mission Control.

We already had our first prototype from our work with CosmoBot. It was a hot mess, but we were on the right track. For this first iteration, we had welded together some metal for the case and wired together several switches and joysticks to control all of CosmoBot's functions. We'd also attached a microphone that the kids clearly loved using.

The second prototype focused on a flexible form factor with just two capabilities: voice input through the microphone and navigation or interactivity using four buttons. We loved this version because you could collapse the buttons into a square shape or line them all up next to each other.

Then we learned about "design for manufacturing." The more pieces you have, the more expensive it is to manufacture and the more chances of something not working. So that led to prototype three, a much more robust design.

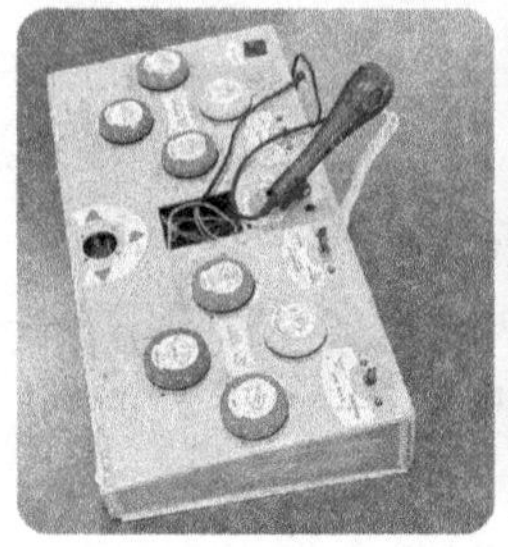
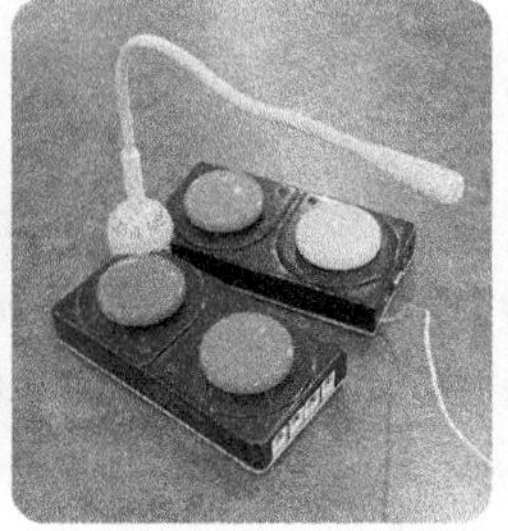
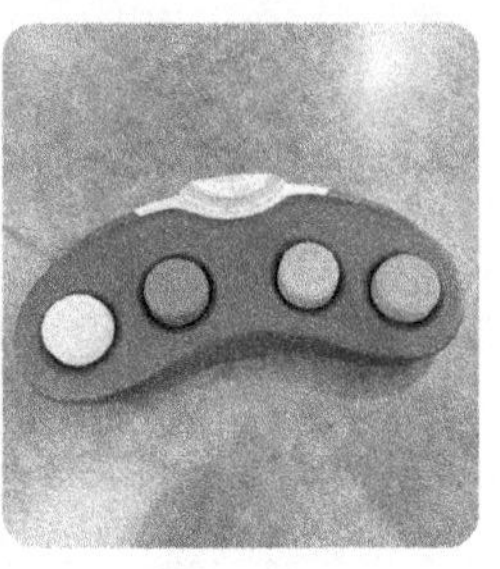

The evolution of Mission Control. (Photo courtesy of AnthroTronix.)

But the real innovation of CLS happened when we tried to interface Mission Control, our alternative keyboard, with existing educational software.

A popular educational software at the time was called Reader Rabbit. In theory, this software taught kids their alphabet and eventually to read. However, my two-year-old, Lindsey, found that she could make it through the software just by pressing the different buttons or by clicking randomly until she hit the right answer. In talking with experts in the education field, we learned that educational software in general was not very educational. It was designed more like a game. In other words, the software had been designed to fit the limitations of the interfaces, like a mouse, instead of being designed to optimize learning.

Software programs designed specifically for kids with disabilities had to work with even simpler interfaces, like the BIGmack button, and so were even more limited in their functionality. Think of an on/off light switch. The button is either pressed or not. It's either right or wrong. Nothing in between. Same with a mouse click. What we needed was a "dimmer switch" button. The more you press, the more something happens. Our final Mission Control had a microphone and four pressure-sensitive buttons. We called these Activators.

But what were we going to do with our Activators? We had already determined that existing software would not leverage the increased functionality we had developed with Mission Control. For that, we needed to develop our own educational software.

SCOPE AND SEQUENCE

We already had a star character for our software, CosmoBot. We knew CosmoBot was appealing to kids of all ages, so we made a virtual CosmoBot character, "Cosmo," for our software. Next, we developed a world for Cosmo to explore. We included a theme song, friends, and lots of cool technology to make shapes and bubbles.

To understand what kids needed to learn, we developed what is called a scope and sequence: basically, a table of contents of the concepts, topics, and materials covered in a particular curriculum. We had never created something like that before, but I knew someone who had: my mother! An entrepreneur herself, she and my father were educators turned entrepreneurs and had run a math curriculum business for years. She pointed us to guidelines published by the National Association for the Education of Young Children, which are used by schools to develop their curriculum. Using those valuable and well-researched guidelines, we developed a complete scope and sequence.

The guidelines included pre-literacy and pre-numeracy goals that were targeted at typically developing preschoolers. We realized that we could adapt our software to serve a wider age range of kids working on these same goals.

For instance, we featured a bubble machine to teach the pre-numeracy goals of "more and less" and "bigger and smaller." Kids who were learning the concept of more would be prompted to press harder on an activator to make Cosmo blow *more* bubbles. In a related activity, the harder the child pressed an activator, the *bigger* the bubble grew.

Student identifies shapes with CosmoBot. (Photo courtesy of AnthroTronix.)

We also provided a set of magnet-based manipulatives and worksheets so teachers and therapists could create a whole lesson plan around any concept that was part of the scope and sequence.

OBSTACLES TO OVERCOME

Once our prototype was refined, we tackled more obstacles to bringing CLS to market. One was funding. We had financed the development of CLS with a combination of government grants from the NIH, the NSF, and the Department of Education. Through our NSF SBIR grant, we now had a new opportunity, as they had just instituted a "super-sized" option. If we raised $1 million, they would match it with half a million.

We returned to our angel investors as well as some new investors, raised the funds, and got the matching grant! So now we had $1.5 million to bring CLS to market. This actually wasn't as much as it seems because it would have to cover software development, manufacturing, and ongoing operations until we had actual sales.

Another obstacle was our company name. AnthroTronix was a great name for a research and development company but a pretty bad one for a consumer-facing company, especially since this was in the aftermath of the post-9/11 anthrax poison scares. Whenever we said AnthroTronix, someone would inevitably stumble over the name, saying, "Anthrax...onics." Sigh. So we decided to spin off a new company, which we named AT KidSystems. "AT" was a nod to AnthroTronix, but more importantly, it was a recognized acronym for assistive technology.

Manufacturing was also a challenge. For our first run, we estimated that we could only afford to manufacture about 500 systems, which was too few to interest most manufacturers. It wasn't until the mid-2000s that Congress invested in US manufacturing, awarding several states with funding to develop capability. That's how we ended up in Shenzhen, the hub of manufacturing about two hours north of Hong Kong. Our experience manufacturing in China was the culmination of developing a product soup to nuts—from beginning to end.

As an inventor, it's rare to see your idea come alive in a completed product. What could be better than that? I'll tell you: going to market and seeing kids use it.

COSMO'S LEARNING SYSTEM DEBUTS

In 2006, AT KidSystems launched CLS at the annual meeting of the Assistive Technology Industry Association. Because CosmoBot was both a virtual character in our software and a physical robot being used for ongoing research, we had a mascot and we put it to work. We brought CosmoBot to several conferences and posted videos on YouTube, which you can probably still find if you search for "CosmoBot and AnthroTronix." CosmoBot participated in the Annual Walk for Autism and was invited to *Wired m*agazine's NextFest showcase.

One of our success stories was working with the Linwood School for children on the autism spectrum. Many of the children attending school there had a hard time using a keyboard and mouse, not only because those devices demand fine motor control but also because they require sustained attention and focus.

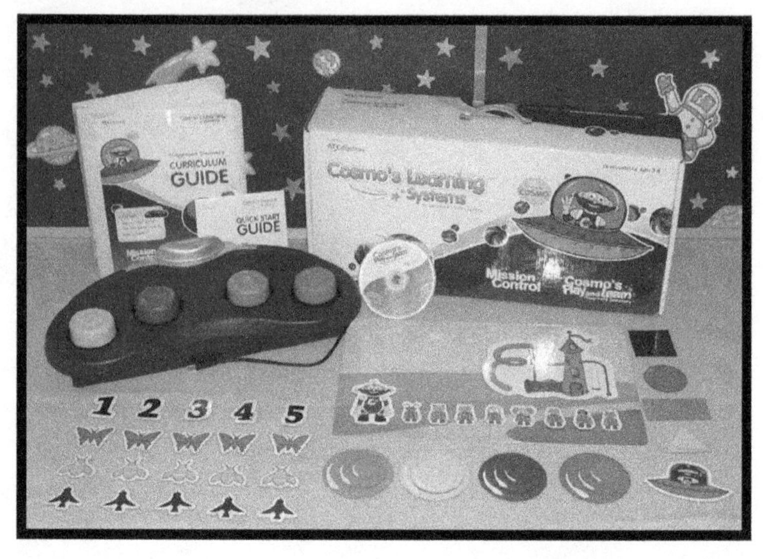

The completed Cosmo's Learning System. (Photo courtesy of AnthroTronix.)

Mission Control, in contrast, was robust and simple to operate. When a child pushed one of the brightly colored activators, they saw clear responses on the screen. Therapists could easily introduce CLS to a kid with behavioral issues who might have trouble sitting in front of the computer. And depending on where the child was developmentally, they could get immediate feedback and progress to more and more complex instructions and concepts.

After our launch of CLS, we were honored with the Tibbetts Award, the top honor for small businesses awarded by the US Small Business Administration. It felt like we were on the road to success!

What We Learned

Cosmo's Learning System was a damn good product.

So why doesn't it still exist? Two words: the iPad.

To put that in context, you have to understand that technology products have a limited life span. The average is less than five years, and that window is getting narrower. For example, a new iPhone model comes out almost every year with features that can render a two- or three-year-old model obsolete.

The iPad presented a truly disruptive shift. It revolutionized how kids and adults interacted with digital content. The first model came out in 2010 and was an overnight success—not just in the general population but also within the disability community. The touchscreen was more accessible than a keyboard or mouse, and it was much more portable and affordable than a desktop.

So we had two choices. Pivot again and reinvent our content for the iPad, or accept the fact that CLS had reached the end of its life cycle. The biggest argument for discontinuing CLS was that our original driver for it was Mission Control. Mission Control offered a simple, powerful way to interact with content, and our software leveraged that interaction in an educational environment. Yes, we could have reworked the software for the iPad, but that would retread innovations we'd already pioneered.

We also had another problem, a totally unexpected obstacle that had little to do with disruptive technology. It arrived in the mail one day, a letter from a lawyer at Hearst Publications. The letter claimed that "Cosmo" violated their trademark for *Cosmopolitan* magazine. This multibillion-dollar conglomerate, one of the world's largest publishers of magazine media, threatened to prevent kids with disabilities from learning with CosmoBot unless we rebranded our product.

Our small assistive technology company did not have the funds to do that.

Hearst agreed we could sell the rest of our existing inventory but could not create any more CLS-branded products.

So that was that.

CosmoBot, the robot, was before its time, though it lived on through the research publications of the Mayo Clinic and other leading institutions. CosmoBot, the software character, had a limited shelf life, but optimizing human–robot interaction became a core competency in our work at AnthroTronix. In fact, we were already leveraging elements of this technology in our work with the military, which you will see in the next chapter.

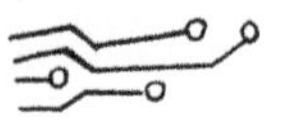

INSPIRATION AND REFLECTION #2
Smart Toys and Beyond: Babies, Puppies, and Segways

The term "social robot" wasn't in wide use when AnthroTronix first developed CosmoBot and brought Cosmo's Learning System to market, but it's a hot topic today as robotics and AI grow ever more present in daily life. They can take on friendly forms (like Sony's pet dog, Aibo) or be less tangible (think Amazon's virtual assistant, Alexa). Either way, if a robot, physical or otherwise, motivates us or elicits an emotion, I consider it to be social.

Although most complex social robots are found in academic research labs, toymakers have led the consumer market in incorporating the technology into their products. One of the earliest examples was Baby Alive (Hasbro), which came on the scene in 1973 as a doll that responded when fed and held. In 2000, Hasbro partnered with iRobot (the maker of the Roomba robotic vacuum cleaner) for the next iteration. My Real Baby boasted even more relationship-building opportunities—she could increase her vocabulary if you taught her new words and suck her thumb when she needed soothing.

A more sophisticated toy offering was the robotic dog Aibo, which has been on and off the market since 1999. The 2018 reboot included adaptable behaviors that let you raise Aibo from a puppy. This social robot could even learn tricks like striking a pose or playing catch—if you were a patient trainer!

While toymakers led the way, other industry innovators have brought social robots to different markets.

Pepper from SoftBank Robotics was one of the first social robot products designed for use in commercial and corporate settings. At forty-seven inches tall, it wasn't as cute as a baby or a puppy, but this humanoid robot recognized faces and engaged in conversations via a touchscreen in its chest. Starting in 2014, SoftBank produced about 27,000 Peppers worldwide. Many of them offered information in airports, greeted customers in banks and office buildings, and provided shopping assistance in stores and malls.

The future of humanoid social robots in these settings is unclear as these environments may be better served through traditional automation (e.g., stocking shelves) and keeping jobs involving customer interface to the real humans. Just because I'm an optimist doesn't mean that I believe robot friends will be everywhere!

Some social robots are less humanoid and more practical. VGo (Vecna Technologies), Temi (Temi), and Bo (BotsAndUs) are great examples. These are being marketed for home use, medical and educational facilities, and also as service robots in public places. Looma (Segway) is a bit different in that it combines the aforementioned functionality with the personal mobility platform of a Segway. So the robot isn't the only one who is mobile—you are too.

Finally, some newer commercial social robots are focusing on the healthcare market—in particular, the world's growing elderly population. In the 2012 movie *Robot and Frank*, "Robot" is based on elder care robots being developed in Japan. Robot can cook, clean, learn new hobbies, play games, and is pretty darn smart. We are a long way from that, but we have some great starts. Mabu (Catalia Health) engages patients, particularly the elderly, as a wellness aide. Mabu can encourage their elderly friend to do some gentle exercises or take a walk, and remind them to do things like take their medication or call a family member.

In the future, social robots will take many forms, including physical and virtual, humanoid and not. They will give form and function to embedded AI software that is only "intelligent" in that it helps us be better humans.

IF you share technology across disciplines,
THEN you can solve more problems.

CHAPTER 3

Sensors for Soldiers

THE STORY OF ACCELEGLOVE

IT'S 2003, AND we're enjoying our first holiday party in the new AnthroTronix space in downtown Silver Spring. Since graduating from the University of Maryland business incubator, we've grown in leaps and bounds, and I'm elated at the sight of new employees and clients getting acquainted. My husband is introducing guests to our bouncy six-month-old, Lindsey, and everyone is clamoring to hold her.

I'm enjoying a glass of wine with our CosmoBot program manager from the NSF, Sally Nerlove, when her eyes widen. "Who is that holding your baby?"

With a burst of alarm, I look around but relax when I see a tough-looking guy, a little older than me, lounging in a chair with his eyes shut. Lindsey is napping in his arms, oblivious to the noise and celebration. "That's Colonel Blitch, our DARPA program manager. His record is so top secret, we're not even allowed to ask what he did during his time with the military."

But in this moment, he's also the perfect analogy for dual-use technology: a fearless defender of the free world who can snuggle a baby without batting an eye. So how did I get to the point where a tough-looking soldier is doubling as my nanny? Let me back up...

Our First Military Contract

In 1999, at the same time we were working on CosmoBot for the NSF, we were invited to bid on a project for another government agency. The Defense Advanced Research Projects Agency (DARPA) is the high-tech research arm of the DoD, tasked with developing new technologies for military use. DARPA had started a new program called Tactical Mobile Robots (TMR)—a far cry from kids with Lego bricks and cuddly mascots. Its goal was to develop robots to use as tools for tactical operations in urban environments and high-risk war scenarios. The program manager was Colonel John Blitch, whose reputation as someone who doesn't suffer fools preceded him. He had a vision for using robots as "team members" during high-risk operations like rescuing hostages or defusing bombs. It was typical for DARPA: a big hairy audacious goal (BHAG) with maybe a 10 percent chance of success. Our task at AnthroTronix was working on the robot controllers.

At the time, 1999, robotics was very much a new technology. A few companies were just starting to manufacture robotics, and their controllers were rudimentary or nonexistent—usually just a computer or, if they were really fancy, a repurposed game controller. We wanted to create a better solution for controlling robots during deployment.

Now, given that we developed our wearable sensor technology to help kids with disabilities, it may sound incongruous that we would bid on a project for the military. But in my mind, it was the same human–technology interface design problem. How do you enable someone with limited ability to have access to the technology they need? In the case of the kids, it was a physical disability that limited their ability to access a computer. In the case of the marines, it was an environmental requirement (e.g., lying in a trench, holding a weapon) limiting their ability to control a robot to go into a building to look for hostages. In both cases, technology should enable capability.

To win the contract, we bought MilSpec (military specification) patrol gloves to replace the cute Mickey Mouse gloves we used with kids. But we embedded some of the same technology: pressure sensors in the fingertips and bend sensors to capture the shape of the hand. We demonstrated that just as soldiers used hand signals to communicate with each other, they could use them to communicate with the robot. These signals could be discrete, like stop or go, or they could be continuous, like a joystick. For instance, a controller could be programmed so that the more you bend your hand forward, the faster forward a robot would move. If you bend back the other way, it could slow, stop, and then move backward.

Knowing Our Value

We demonstrated our concept for a team at a large DoD contractor that Colonel Blitch had funded to lead the effort. The team loved our concept for an advanced controller and contracted us to develop it further. This was our first experience with a large DoD contractor, and it was intimidating and a bit disconcerting. For example, we were expected to deliver a product that would take a huge investment of time and talent to achieve *before* getting paid. This was a fairly common practice in government funding, in part because typical government bureaucracy couldn't match the pace of the demanding deadlines of DARPA projects.

Companies work under these terms all the time, as the risk of not getting paid by the government is low. However, as a startup, AnthroTronix neither had the money nor the experience to manage cash flow in this situation, so I just said, "No. We can't forward-fund the government. But you can, so pay us if you want to work."

Normally this would get a subcontractor (us) fired, but the great thing about working in high tech is that sometimes there isn't an easy second choice. Two days later, they paid up and we learned two valuable lessons. The first was to recognize our value. The second was to structure every contract with an up-front payment!

Part of our value was a unique advantage we had—my co-founder Jack was a former marine. His military experience was crucial to many aspects of our work with DARPA, not the least of which was how to communicate with our program manager.

The Next Hurdle: Our Program Manager

DARPA programs can be very high intensity. Program managers like Colonel Blitch are typically hired for just two years, so they have a short time to show success. We had only one month to strategize and put together a presentation for him. Blitch hadn't been impressed by the lack of progress at the first quarterly progress review (QPR). He'd already cut two teams from the program, and we could be next.

The day of the presentation came, and to say that I was nervous would be an understatement. I was thirty-three years old and had left a stable job in academia to build a high-tech startup. We were running the company on a shoestring budget, and blowing this could mean the difference between making payroll and not. If Colonel Blitch didn't like what we had to say, we were out. That would spell disaster for our fledgling company.

On the day of the presentation, I stood before an audience of about fifty in a cramped room, seeing Colonel Blitch for the first time. By this point, I'd heard the rumors regarding his military experience as part of a Special Missions Unit. Google it. It's intimidating.

I donned a glove with a few sensors strapped to it and started controlling a toy robot in front of me. I compared the gestural interfaces we were proposing to implement with soldiers with what we were doing with kids with disabilities.

This was a big leap, showing a kid's toy and talking about military tactics, but I explained our philosophy on human–robot interaction: "The humans shouldn't have to learn the robot's language; the robots should learn ours. Soldiers use hand signals to communicate with each other, so why not use hand signals with the robots too?"

I'll never forget seeing Colonel Blitch's face light up for the first time! "The beauty of your glove is that you can capture the hand signal digitally and then transmit it to a robot or to another soldier."

"Exactly, sir." (I may have stumbled on the "sir," only remembering at the last moment how Jack said I should address him!)

The colonel immediately saw that this could solve some sticky tactical problems. "Your glove could enable communication even when two soldiers can't see each other, which is a huge advantage when patrolling in the woods or around obstacles." We agreed!

Colonel Blitch was intrigued. He went on to say, "But I don't want to take my hands off the weapon." He was frustrated with the developers who expected soldiers to sling their weapons and then pull out a bulky game controller whenever they wanted the robot to do something.

At this point, Jack chimed in with a ready answer. "Unfortunately, in some cases, the robot operator may need to be a protected military occupational specialty, like a radio operator. The level of fidelity might demand slinging the weapon and focusing on robot control. But we understand your position and have a potential solution that should work for most field operations."

We then showed a concept drawing of a soldier lying in a trench holding their weapon at the ready but pressing an instrumented glove's fingertips against the barrel of the gun to control the robot to go into a building.

Colonel Blitch gave us his blessing, and we were off and running to develop an instrumented glove for robot control.

Developing the Prototype

Our first working prototype used flex sensors along the fingers that changed their resistance as they bent. With constant current, we could measure the change in voltage as the fingers bent the sensors and use that signal to control the robot.

We also made homemade pressure sensors for the fingertips, using two tinfoil sheets with an insulator between them. Pressing against the sensor caused the capacitance (the ability to store electric charge) to increase and the measured voltage to decrease.

Between the flex sensors and the force sensors, we now had a glove that gave the wearer direct control, the way a joystick does. The wearer could control a robot using finger movements or by pressing their fingertips against a hard surface and could capture a command like "halt" with a particular shape of the hand.

Putting It to Work

We had a problem: to test our solution, we needed a robot, and no one had one to spare. Most early robot developers, like iRobot and Boston Dynamics, were either using specialized one-off robots for demonstrations or going into very limited production. So we had to develop our own solution. Our light bulb moment was that we didn't need a physical robot to test our controller—we could accomplish our goal with a simulation. To create our virtual robot, we used the Torque 3D computer game engine, a free, open-source, cross-platform toolkit.

Our experience with Torque gave us a head start on many of the software skills we used as a company later on. Building a simulation was a big investment for a small company, but it ended up paying off in spades, becoming the basis for our first glove product after the TMR program ended.

But I'm getting ahead of myself. Back to 2001.

The TMR program was in full swing. The January 2001 issue of *National Defense* magazine featured our gloves in a piece on advances in human-computer interaction.[9] JB—as we now thought of our gruff-voiced, impatient, and extremely dedicated program manager—explained how a Tactical Mobile Robot would be controlled. "The primary interface is a pair of gloves that act as regular protection until a button is pressed. They then become gesture recognizers that control the robot." He adds that these gestures are better than speech recognition, which can be misinterpreted by the robot or detected by enemy forces.

Interestingly, JB also used the article to advocate for a new military operational specialty for robot operations. He pointed out that controlling a robot can degrade a soldier's primary skills.

Clearly, our shared vision, discussed at that fateful first meeting eighteen months prior, had come to fruition.

Shifting Focus to the AcceleGlove

TMR came to an abrupt conclusion in September 2001, when the events of 9/11 caused the military to shift focus. At that point, the most mature piece of TMR technology was the PackBot, a ground robot developed by iRobot. PackBots were deployed to Afghanistan to help with explosive ordnance disposal. Many of the personnel, including JB, were deployed as well.

With the close of the TMR program, we were left with our instrumentation capabilities and new simulation/software capabilities. The

9 Duffy Baker, "User-Friendly Machines Help Boost Performance in Robots," *National Defense*, (January 2001): 37–38.

question now was, what should we do with them? Luckily, this technology wasn't only useful for the military. It was dual-use, with both DoD and civilian applications. Now that DARPA funding had ended, it was time to explore the civilian side.

To develop our glove for TMR, we had examined how other innovators were meeting the challenge of measuring hand shape and movement. We'd run into several projects related to sign language. It seemed that every few years, a student project would win an award for "translating" sign language. Usually, the solution captured the twenty-six static hand positions for the American Sign Language alphabet—a great student assignment, though it barely scratched the surface of capturing the dynamic movements of sign language.

But one of these winning students, José Hernández-Rebollar, was using an innovative way to capture more of the richness of sign language by using accelerometers. Accelerometers measure the orientation of the hand with respect to gravity—essentially, more complex tilt sensors than what we had used in developing the JesterBot controllers.

José, an inventor from Mexico, was finishing his graduate degree at George Washington University, only a few miles from our office in Silver Spring. He was working with Dr. Corinne Vinopool, the founder and CEO of the Institute for Disabilities Research and Training (IDRT). IDRT's mission is to improve the lives of people who are deaf and hard of hearing, and the institute was interested in using José's glove to teach people basic sign language. IDRT believed there was a market; at the time, Rosetta Stone said that sign language was their second most requested language (I believe Mandarin was the first), but IDRT didn't have our depth of hardware and software product development experience. We loved the idea of partnering with them and knew our experience in gesture recognition for command and control of robots and communication between soldiers could apply to this new project.

We formed a partnership with IDRT to develop the AcceleGlove data glove. We licensed José's patent for using certain sensors for gesture capture and combined it with our patent on using gestures to control computer-based assets. We used our experience in simulation software development to create an open-source software developer's kit so anyone could capture whatever movements they wanted and map them to computer-based assets.

As you may know, the concept of open source is that software is released under a license in which the copyright holder grants users the rights to use, study, change, and distribute the software and its source code to anyone and for any purpose.[10] By the early 2000s, the open-source movement was gaining traction, thanks in large part to innovator Linus Torvalds. When Torvalds created the Linux operating system, he was frustrated with the proprietary nature of operating systems and felt progress would happen quicker if people were able to crowdsource capabilities. So he released it under an open-source license. Linux became hugely popular in the tech world, giving rise to the Android in 2007, and open source is now a common practice among developers.

We also combined forces in manufacturing. José had contacts in his home country of Mexico that could make the textile part of the glove. For the electronics, our manufacturing options were much closer to home than they had been when we developed Mission Control. Though it was only a few years later, investment in manufacturing in the United States had surged. In response to a 2005 National Academies of Science report on the failing technical leadership in the United States, George W. Bush announced the American Competitiveness Act to increase US capabilities in manufacturing. In 2008, a new manufacturing facility, the Southeastern

---

10 "Open-Source Software," Wikipedia, last modified March 2, 2022, https://en.wikipedia.org/wiki/Open-source_software.

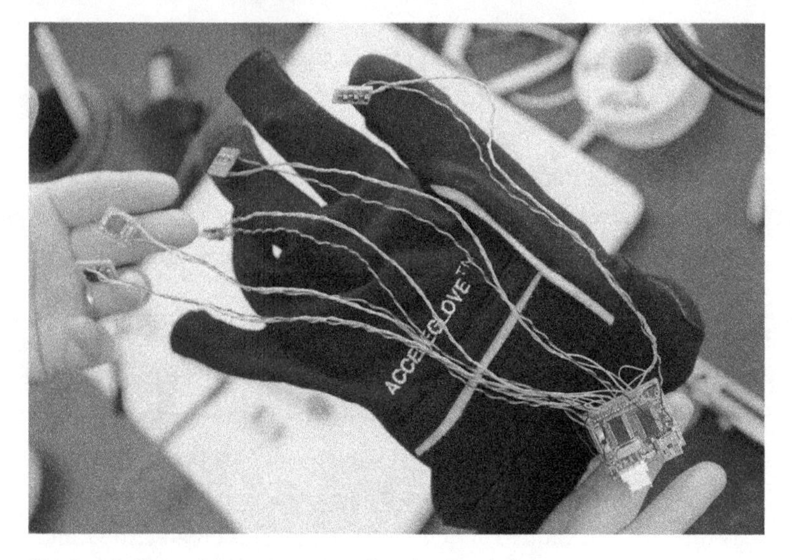

The AcceleGlove. (Photo Courtesy of AnthroTronix.)

Institute of Manufacturing and Technology (SIMT), opened in South Carolina. SIMT worked with us to develop the electronics and produced the electronic wiring harnesses. We then assembled the gloves at AnthroTronix.

The Market for Data Gloves

In 2009, we released the AcceleGlove to much acclaim—MIT's *Technology Review* heralded its many innovations in an article titled "Open Source Data Glove: AcceleGlove Can Be Programmed for Many Applications."[11] This title may seem to state the obvious, but at the

---

11 Kristina Grifantini, "Open-Source Data Glove: AcceleGlove Can Be Programmed for Many Applications," *MIT Technology Review*, June 23, 2009, https://www.technologyreview.com/2009/06/23/212253/open-source-data-glove-2/.

time, hardware products typically didn't provide that kind of open-source software access to developers. Our decision to make the developer's kit open source meant that users could adapt the software for their own purposes even beyond the initial capabilities of the glove or software. We wanted to go beyond sign language and allow users access to gesture capture to control any computer or video game. Or to create music or art. Or applications we hadn't thought of.

The article also pointed out that "other gloves...normally cost $1,000 to $5,000, but the AcceleGlove costs just $499." One of the key developments in the industry that made the glove affordable was the consumer electronics revolution. More and more off-the-shelf sensors were becoming available, meaning anyone could go to RadioShack or order sensors online, including accelerometers. Accelerometer technology enabled us to capture a wealth of hand positions cheaply.

At $499, the AcceleGlove was half the price of other data gloves on the market, and early adopters loved it! But the same trend that had enabled us to bring it to market affordably continued. After only about six months, new sensors entered the market that were faster and cheaper than the ones we had used. In terms of consumer electronics, our glove had become obsolete. Most customers, particularly the early adopters, wanted the latest and greatest.

In the end, we sold only about 150 gloves, mostly to students at universities using them for research purposes or to developers and inventors for projects targeting the entertainment industry. IDRT also discontinued marketing the glove but continues to this day doing their good work, providing resources and computer-based technology to the deaf community.

Lasting Impact

Interestingly, we continued glove development work for another decade. Because we had become the experts, everyone from government agencies to individual customers asked us to make custom gloves. For example, we created a version of our data glove for the Second Life 3D virtual world that enabled veterans using wheelchairs to rock-climb. For another client, we made game controller gloves for stroke rehabilitation patients. They could work on fine motor control by controlling a game with their fingers. To help develop training protocols for surgeons, we measured their hand movements while using surgical simulators. Yet another application involved capturing military police hand signals at checkpoints so they could communicate warnings silently.

Coming full circle, the final version of our data glove, called the nuGlove,[12] was used in another DARPA contract called DROID AGENT as an input to facilitate machine learning of medical procedures. The device supported an automated training and coaching system for combat medics and navy corpsmen. The application of a tourniquet to an arm was the initial use case. All of these one-off demonstrations of our glove helped our clients pitch a vision of the future to their government or private investors.

But what about the outcome of the TMR program? Colonel Blitch's faith in our abilities helped launch our company, and we now had lots of other projects to keep us busy. And whenever JB is in Silver Spring, he is a welcome guest at our parties.

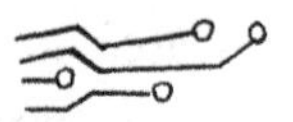

INSPIRATION AND REFLECTION #3
Movement-Based Interfaces

I've told you about kids and soldiers using natural movements, or gestures, to interface with robots and computers. Our own Accele-

12 "Projects Using NuGlove," AnthroTronix.com, last modified January 30, 2018, https://www.anthrotronix.com/nuglove-projects/.

Glove was the culmination of our work; it digitized the motions of fingers, hands, and arms to control a robot or computer, communicate, or create new content. This idea of motion capture is not new, and our work was inspired by early instrumented gloves that actually began decades ago in the gaming industry.

In 1989, Nintendo introduced the Power Glove, which was followed up a few years later with Virtual Technologies' CyberGlove. Gesture-based game controllers remained in obscurity until Nintendo's Wii controller brought this idea to consumers in 2006. The Wii controller was a handheld device that sent position and acceleration data to the system as you moved it, which then used that data to move a corresponding object (e.g., a dancing human avatar, a racket in a tennis game). Like the AcceleGlove, the Wii controller used accelerometers and other sensors to track its orientation.

The use of Wii's gesture-based controller peaked around 2010. This was about the same time as the emergence of Microsoft's Xbox Kinect. The Kinect motion-capture system was camera based, which had the advantage that the gamer didn't need to wear or carry a controller while dancing or playing a sport. The Wii and Xbox gestural controllers had some commercial success with casual gamers. Most gamers never abandoned their traditional controllers, and both companies discontinued these motion-capture interfaces by 2017.

Another reason for the end of the console-based gestural interfaces was that in 2016, Reality Lab released the first version of the Oculus. Oculus virtual reality headsets use integrated cameras to track hand movements paired with inertial sensors in the handheld controllers. With this system, you were no longer tied to a game console and you had movement-based control that was meaningful. One of the most popular VR games, *Beat Saber*, is a great combination of gaming skill and exercise as you use virtual sabers to slice fast-approaching blocks to a musical beat. Or our family VR favorite: *Star Wars*. My family still laughs at me for falling over when I leaned on the virtual cockpit of a spaceship I was flying!

The future will bring increasingly more sophisticated real-world applications of gesture recognition, such as in the surgical room, where touchless control would maintain the sterile environment. Currently, GestSure has a camera-based system where doctors can use simple hand gestures to access MRI, CT, and other imagery without breaking the sterile field. In addition, gesture-based interaction is already entering common use in some consumer products like car dashboards (where drivers can control climate with a wave of the hand) and restroom sinks and hand dryers.

Meanwhile, glove-based motion capture, or "mocap" as it's called in the film industry, has continued to develop for industries that need the high-resolution finger and hand movement of an instrumented glove. One current example is StretchSense's MoCap Pro, which primarily focuses on the animation industry and is used to capture an actor's movements to develop a computer-generated imagery character. Another glove still on the market as of 2021 is the 5DT Data Glove Ultra; it captures finger and hand movements to manipulate virtual objects and navigate through high-fidelity training simulators.

Glove-based interfaces are also starting to incorporate haptic technology (touch and force feedback) to better immerse the user in the experience. Haptic technologies have been evolving for decades, and it's only a matter of time before they become embedded. HatpX Gloves have already started providing force feedback for the high-end training market. Beyond gloves, the Emerge Wave-1 is a tabletop device that uses ultrasonic waves to create a social-tactile experience in VR that can bring together a community of users.

Whether for controlling complex robotic systems, smart appliances, medical devices, or exploring mixed-reality environments like the "metaverse" (more on that later in the book), gestural interfaces will soon be indispensable for capturing our natural movements, digitizing them, and using them to expand our capabilities.

IF you look beyond the invention,
THEN you can impact the community.

CHAPTER 4

Virtual Reality in Space

THE STORY OF THE 3D SPACE EXPERIMENT

THE KC-135 AIRPLANE, fondly called the Vomit Comet, takes off from Ellington Field near the Johnson Space Center in Texas with only a handful of people on board. The windows are completely covered and most seats have been cleared to make room for exercise machines, heavy padding, and lab equipment. We are all wearing drab green jumpsuits with little white bags sticking out of our pockets like handkerchiefs.

After thirty minutes, the plane reaches level flight over the Gulf of Mexico. We scramble up to get to our workstations. I secure the VR headset on my test subject's head and then strap myself in on the floor next to him. At the next workstation, my fellow researcher Gilles does the same. The flight engineer gives us the one-minute warning, and seeing Gilles take slow, steady breaths reminds me to do the same.

Cori and Gilles collecting data in the Vomit Comet in Houston, Texas. (Photo Courtesy of NASA.)

I start to feel heavy, as if I'm being pressed into the floor, kind of like a gravitron centrifuge ride at the amusement park. Twenty seconds later, my whole body starts to rise except for my middle, which is still strapped in. I unbuckle myself and push off into the air, my hair flying around me. A piece of it floats into Gilles's face and I pull it back, laughing. I want to spread my arms and do a somersault, but we're here to work. My subject is still strapped in and watching the video stream. I check the readings as my equipment records his eye movements, and after about twenty-five seconds, I start to sink back down. The drop makes me queasy, but I give Gilles the thumbs-up as I strap myself back on the floor.

I still can't believe we are here. If Gilles hadn't gone to bat for me with NASA, the experiment might not have happened at all. We watch the flight engineer for our next one-minute warning, eager for another mini-trip to zero gravity. This may only be simulating the weightlessness of spaceflight, but it's the culmination of a life-long dream, and it's only happening thanks to dogged persistence, international collaboration, and of course, *Star Trek*. Let me back up...

Calling All Trekkies

In fifth grade, I was voted most likely to go to Mars. My obsession with space actually started before that, when my second grade teacher designed a whole curriculum around *Star Trek* characters. To become a communications officer like Uhura, you had to learn to count to ten in two languages besides English. To become a navigator like Sulu, you made a map from your home to the school. To become a science officer like Spock, you had to do a lot of math problems. The first person to reach all of these goals became the captain.

The race for James Kirk's chair came down to me and one other girl. We were on the Spock challenge and I got *one* math problem wrong. I was devastated until I learned the other girl had gotten *two* wrong. I became our class's first captain! I even wore the pin for my school picture that year.

From that point on, I watched every *Star Trek* rerun. But what fascinated me most wasn't the characters—it was the technology. I wanted a tricorder that could diagnose and treat almost anything instantly. I wanted to visit a Holodeck and have adventures. I wanted a replicator that delivered my favorite foods instantly. I told my parents I wanted to be an astronaut and invent space technology.

Proud captain, with pin.

In college, although I retained my passion for spaceflight, I explored different paths to get there. I made up my own major at Swarthmore College, biopsychology and mathematics. Basically, I threw together all the classes I liked and gave it a name. I discovered a love of scientific research, and various summer jobs included working in an infant perception lab (watching babies look at objects) and working in a genetics lab (watching fruit flies mate). But I dreamed of working at the Ashton Graybiel Spatial Orientation Laboratory at Brandeis University, a place dedicated to studying how humans could live and work in space. I bugged them repeatedly for a job, which I got the summer after my junior year.

My summer at the Graybiel Lab was transformative. I experienced weightlessness for the first time as a subject for the lab's experiments on the KC-135, and though my research would continue to take me back to the Vomit Comet over the years, I never took for granted the wonder—or the queasiness—of working in zero gravity.

I also discovered a love of neuroscience and a fascination for how the brain interprets signals from our senses, which then forms our perceptions of the world. Take, for instance, the oculo brachial illusion. This illusion tricks your brain into thinking an object attached to your finger has moved even if it hasn't, by sending false muscle movement signals to the brain. With a little effort, you can try this illusion yourself. (1) Brace your bent arm so that it can't move. (2) Attach a small light (like a Christmas LED light) to your fingertip. (3) Apply mechanical vibration (like a muscle massager) to the brachial tendon—basically, the inner part of your upper arm. Oh, and don't forget to turn out the lights before you begin so the only thing visible is the small LED light.

The vibration sends motor signals to the brain that the tendon is lengthening. When you watch the small light while this vibration is occurring, you perceive that your arm is extending and you "see" the light on your fingertip moving. All while your arm remains braced, and the light and your eyes remain fixed.

One hypothesis is that your brain thinks your eye is moving, which seems plausible since you already think your arm is moving. The second, more exciting hypothesis is that your brain thinks the light is actually moving. This hypothesis is more radical because it says that an object's true location as interpreted by the image on the retina of your eye can be reinterpreted or remapped by the brain based on false muscle information.

To test these hypotheses, we extended the experiment and put lights on fingertips of both arms to see if the lights would appear to move away from each other when the arms were vibrated. They did! Therefore, the first hypothesis couldn't be correct because if the brain thought the eye was moving, it would also interpret the distance between the lights as remaining consistent. But even though the lights' positions on the retina hadn't changed, the brain thought that they had.

I learned a lesson that was directly applicable to spaceflight research: As sensory inputs changed in unexpected ways, the brain had to "guess" how to interpret the signals, which resulted in a misperception of our world. During spaceflight, sensory inputs definitely changed and we had only scratched the surface on what that would mean for human performance!

Brains in Space

When we first began sending humans to space, we didn't know much about what would happen to the human body. One reason dogs and monkeys were the first astronauts was that scientists didn't even know if hearts would beat or if lungs would inflate. It turned out that the effects of zero gravity are much more insidious. Fluid flows to the upper body, our bones start to demineralize, our vision changes, and our balance changes dramatically. For people to successfully live and work in space, more research was needed to understand how the human body, especially the brain, functions in space.

One crucial brain function that works differently in space is the vestibular system, or balance. To quickly get in touch with your own vestibular system, hold a piece of paper with writing on it in front of your face. Now shake it. It should get blurry. Next, hold the paper still and shake your head. You should see a difference. When you shake your head, the writing shouldn't look as blurry as when you shake the paper. That's the difference between using your visual system only (shaking the paper) and using the visual and vestibular systems together. In fact, if your visual–vestibular system didn't work properly, you couldn't walk and focus on where you were going. You would have to stop in order to see what was ahead.

The vestibular system is part of our inner ear. It senses both linear and angular motion in m/s^2. Linear motion is what you might feel in a moving train or car; angular is motion around a fixed point, like spinning in a chair. Those same sensors send messages to your brain regarding your position with respect to gravity. Remember that gravity is just the amount of acceleration: 9.8 m/s^2, which is pulling us to the center of the planet (where gravity is equal to zero). So when you jump, you fall back toward the center of the Earth at 9.8 m/s^2. This is why the vestibular system's sensors also know if your head is tilted ninety degrees toward the ground or just fifteen degrees. But in space, when you tilt your head, it doesn't change with respect to gravity, so these sensors don't signal the brain that it has moved. The brain needs to recalibrate based on other sensory signals, such as neck muscles and visual cues.

Astronauts learn to function despite this disorienting recalibration of sensory inputs. They would generally take medication to prevent motion sickness for the first few days of a mission until they adapted. But spaceflight missions at that time were relatively short, generally less than two weeks. No one yet knew how longer flights would affect the ability to live and work in space.

I wanted to know how the changing inputs to the brain in outer space could affect our perception of objects and how we interact with them. I and many others thought that understanding what happens to the brain in outer space would not only enable human space flight but also contribute to our overall understanding of how the brain works. Happily for me, Congress declared the 1990s the Decade of the Brain, "to enhance public awareness of the benefits to be derived from brain research." Research into neuroscience exploded.

I was lucky enough to be part of that research after I graduated from Swarthmore. First studying the effects of gravity on the visual–vestibular system with Gilles Clément at the Centre national de la recherche scientifique (CNRS) in Paris during a gap year, then as a doctoral student of neuroscience at MIT, where I joined the Man-Vehicle Laboratory in the Center for Space Research. (The MVL had a long-overdue name change almost twenty years later, renamed the Human Systems Laboratory in 2018.)

International Collaboration

Working with Gilles during my gap year was another life-changing experience. Not just because I was doing neuroscience research, but also because working at the CNRS was my introduction to the tremendous international collaboration behind spaceflight and space research.

Keep in mind this was 1989, and America and Russia had not yet worked together on the International Space Station. The French space agency was simultaneously collaborating with the United States on a space shuttle experiment and with Russia on a space station experiment. Jean-Loup Chrétien, a French cosmonaut, had just spent six months on the Mir space station with a Russian team, and his stories about living and working in space informed much of the science that Gilles and I eventually proposed.

Working in France was also the first time I seriously contemplated the career of astronaut. I realized that astronauts are not just people with a passion for space and adventure, or just excellent engineers and scientists who landed a cool job, but also ambassadors for international peace. I wanted to be part of that and sent in my first application to NASA's astronaut program in 1990.

They didn't tend to welcome grad students into their ranks, so a rejection wasn't a big surprise. But as I continued my career in academia and then with AnthroTronix, I reapplied every time NASA announced a new selection, figuring it was at least worth a shot.

In 1998, the United States, in partnership with Europe, Russia, Canada, and Japan, began construction of the International Space Station, which would be quite a bit bigger than Mir and result in a large, permanent presence in space. Scientists and engineers throughout the neuroscience community shared concerns about how alternate sensory inputs changed what the brain was perceiving and what astronauts were "seeing"—much as I'd discussed in my work on the oculo brachial illusion—and how this would affect performance while living and working in space.

In fact, there were anecdotal reports from the astronauts, including Jean-Loup, that they misjudged distances as they moved around. Yet they were still required to do many tasks that required eye–hand coordination, from navigating inside the station to operating a robot arm outside. We needed more science before proposing potential solutions.

Making the Case for 3D Space

Based on data we'd published from our one subject on Mir, Gilles and I proposed the Mental Representation of Spatial Cues during Spaceflight experiment—3D Space, for short. By this point, we had published multiple scientific articles and written a book chapter on visual–vestibular interaction in space.[13] I had started AnthroTronix and had a team that could develop the technology needed for 3D Space. My dreams of a holodeck were not quite on the books, but I had some ideas for using virtual reality technology to help understand what happens to astronauts in space. At first, NASA and the French space agency (Centre national d'études spatiales, or CNES) loved it.

Then NASA narrowed their focus. They shifted all resources to the complex engineering task of building the ISS. Many US-based researchers, including me, received letters ceasing funding for our spaceflight projects. However, Gilles and his colleagues at CNES went to NASA and said, "We can't fly this experiment without Dr. Lathan. Her team is developing the software. We can't do the science without the software."

---

13 C. E. Lathan and G. Clément, "Response of the Neurovestibular System to Spaceflight," in *Fundamentals of Space Life Sciences*, ed. S. Churchill (Malabar, FL: Krieger Publishing, 1997), 65–82.

My funding was back on! This international collaboration saved 3D Space.

Pursuing the Dream

Meanwhile, I continued to apply to NASA. In 2003, just months after the Space Shuttle *Columbia* disaster took the lives of seven crew members, I was invited to interview for a spot in the astronaut training program.

My husband and I had just adopted a baby. AnthroTronix was thriving and my projects were engaging me on every level. And the tragedy of *Columbia*, with its toll on the country and the space program, highlighted the very real dangers of the job. But the opportunity to do meaningful work in space was equally compelling. I took the interview.

My application essay included a fragment of a poem written by Robert Frost,[14] which captures my passion for the space program and why I would say yes in a hot minute if they would have me. This essay (including the poem) was read aloud to the committee before I went in for my interview.

But yield who will to their separation,
My object in living is to unite
My avocation and my vocation
As my two eyes make one in sight.
Only when love and need are one,
And work is play for mortal stakes
Is the deed ever really done
For heaven and the future's sakes.

---

14 Please see copyright page for license agreements.

This is the last stanza of "Two Tramps in Mud Time," and I first heard it as a child. My parents informed me that this was my family's maxim, or fundamental truth. It still encapsulates everything I feel about my mission in life and why I'm here. We have a moral obligation to ourselves and others as we make the hard choices needed to maintain balance. Knowing that those lines were read to the NASA selection committee just before my interview grounded me in my purpose as I walked in the door.

The interview was an experience that I will never forget. You sit at the table with a dozen or so astronauts who basically decide whether they could stand spending six months locked in a room with you! Our conversation ranged from poetry to robots to hang-gliding. I would happily have spent time in space with all of them.

In addition to the one-hour interview, there were spacewalking simulations, psychological testing, and even a barbecue social with moments of scrutiny. Inviting our family members was "optional," but woe betide the interviewee who showed up alone. Luckily, I was there with my family already as I was still breastfeeding my three-month-old. I hadn't been sure I could manage a week of interviews and testing in Houston with a baby, but thanks to astronaut Cady Coleman, whom I knew from mutual friends at MIT, I had a place to stay and a loaner car with a car seat.

We finalists also spent the week being poked and prodded with almost every medical test imaginable, including a sigmoidoscopy and a mammogram—which, by the way, is not valid when you're breastfeeding. In the end, I was medically disqualified because my eyes didn't quite pass the vision acuity test. At the time, they did not allow Lasik, though a few years later they followed military protocol and allowed vision correction procedures. But by then, I was consumed with our 3D Space experiment and other projects that were in full gear.

How 3D Space Worked

The 3D Space experiment represented the first use of virtual reality technology on the International Space Station. For the hardware, we used the eMagin headset, winner of the 2006 Consumer Electronics Show Innovation Award. It was lightweight but still reasonably immersive for what our astronauts would be doing. We also flew a Wacom digital drawing tablet. We clearly had an eye for good technology because both devices are still on the market today, albeit ours were several generations older.

One of the challenges for AnthroTronix was developing software for an operating system that was already outdated and that we couldn't alter in any way. We couldn't even load the most up-to-date drivers for the headset and tablet. To meet specifications, we had to develop software from scratch that could be installed on the existing flight-approved IBM Thinkpad running Windows 2000.

Our goal was to investigate the effects of exposure to microgravity on the mental representation of spatial cues by astronauts during and after spaceflight. So what did that actually involve?

We gave astronauts three different tasks. For the first, the subject would put the digitizing tablet on their lap, *close their eyes*, and then either write a word (e.g., "microgravity") both horizontally (left to right) and vertically (top to bottom), or draw a geometrical object (e.g., a square or a cube).

For the second task, they wore a head-mounted virtual reality display (or VR headset). The headset displayed an object or a geometric illusion and the astronaut was asked to adjust it. For example, one image was a misshapen cube. The astronaut had to adjust the image so that it appeared as a perfect cube. One of the geometrical illusions included was an inverted *T*. For this, the astronaut was asked to adjust one line in the image so that the two lines appear the same length.

For the third task, a static visual scene was displayed in the VR headset, and the subject was asked to estimate distances between objects and landmarks in the scene.

We measured the astronauts' performance on those tasks while they were in space, on land, and during short periods of free fall in the KC-135 aircraft. Comparing the results helped us understand how spaceflight affected our subjects' ability to interpret spatial cues—their perception of horizontal and vertical lines, object depth, and distance.

THE VOMIT COMET

It's very expensive and rare to actually run experiments in space, but it is possible to have limited exposure to weightlessness, sometimes erroneously (but conveniently) called zero gravity, or 0g. Gravity is still there; you are actually just temporarily in free fall. In a plane. Thirty thousand feet in the air.

To understand why the experience is so stomach churning, you have to picture our flight path. The KC-135 flies a series of forty parabolas. As it accelerates up the first slope, you feel pressed against the floor or seat, experiencing the equivalent of double gravity. At the top of the slope, you are in free fall and feel weightless. That's when we researchers usually unstrap and do our work. Then, as the plane accelerates out of the free fall (so you don't hit the ground), you experience double gravity again. Throughout this, your stomach is rising and falling at every transition from 2g to 0g and back again. Most people get sick at some point, and I was no exception. But that's what the white bag in your flight suit pocket is for!

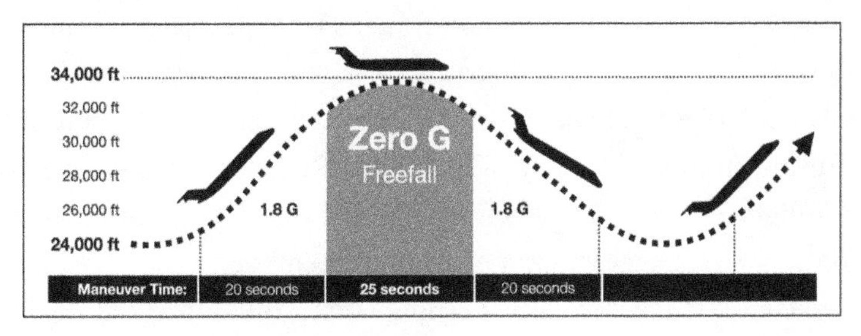

How parabolic flight works.

OFF-WORLD SUBJECTS AND SOCIAL MEDIA

As you might imagine, it's also a bit complicated collecting the data for experiments occurring in outer space. But the astronauts we worked with were not just subjects—they were co-investigators. They ran the experiments themselves on the station, and we spent several days with each one, pre- and post-flight, for data collection. Each astronaut was personally invested in understanding the technology as well as the requirements for the science. They also understood that the taxpayers were paying a lot for this research.

Every crew to the ISS is called an Expedition, with Expedition 1 in October 2000 being the first. The 3D Space experiment launched with Expedition 17 in 2008 and remained through Expedition 27 in 2011. Over this three-year period, we had eight subjects, seven men and one woman.

Because the space research community is small, I knew many of our subjects already. I'd worked with one of our test subjects, Canadian astronaut Bob Thirsk, for my master's thesis. Mike Barratt, a flight surgeon from NASA who was selected for the astronaut corp in 2000, was a longtime friend from International Space University, which I'd attended in Japan the summer of 1992. One of our last subjects, Cady Coleman, had been my gracious host during my NASA interview. It felt incredibly special to be one of the first to welcome her home when she landed after five months in space—and then gather every piece of data from her that I could!

Unlike with most science experiments where subjects are anonymous, the astronauts' participation was public information. Their daily schedules were posted on space-related websites. For example, a European Space Agency's Operations Report from 2009 stated, "ISS Flight Engineer Tim Kopra successfully performed his final session of the 3D Space experiment on 2 September."

Italian astronaut Paolo Nespoli, from the European Space Agency, posted a photo on Flickr with the caption, "Here I am car-

Astronaut Cady Coleman completing the 3D Space experiment. (Photo courtesy of NASA.)

rying out ESA's 3D Space experiment. This experiment is about how changes in visual perception affect motor control. I have to do several different tests wearing virtual reality goggles and write or draw on an electronic tablet. It's all in the name of science!"[15]

One picture that sums up the collaborative spirit of 3D Space was our 3D Space patch, which our subject, Koichi Wakata, flew in space in 2010. He took a beautiful picture from the ISS Japanese Experiment Module, called KIBO, which means "hope." The patch was thoughtfully designed with the ISS and a 3D cube drawing as the backdrop. Layered on it are an American flag and a French flag, anchored with two small stars representing Gilles and me, as prin-

15 Paolo Nespoli, ESA/NASA. 2011, digital image, Available from Flickr Commons, www.flickr.com/photos/magisstra/5363766102/ (accessed March 29, 2022).

The 3D Space patch floating in the window of ISS. (Photo courtesy of Koichi Wakata.)

cipal investigators. Eight small stars float in space representing our subjects.

WHAT WE LEARNED

Over a three-year period, our team collected pre-, in-, and post-flight data on eight International Space Station crew members. Here are a few things we learned about in-flight perceptions versus typical perceptions on Earth:[16,17]

16 Gilles Clément, Anna Skinner, and Corinna Lathan. "Distance and size perception in astronauts during long-duration spaceflight," *Life* 3, no. 4 (2013): 524–537.

17 Gilles Clément, Anna Skinner, Ghislaine Richard, and Corinna Lathan. "Geometric illusions in astronauts during long-duration spaceflight," *Neuroreport* 23, no. 15 (2012): 894–899.

1. In space, a cube is not a cube. Instead, subjects see the dimensions such that the "cube" is wider and deeper than it is tall—more of a 3D rectangle.
2. This "shortening" of the vertical also occurs with objects that the subjects draw with their eyes shut, indicating that both perceptual and motor changes are occurring. Objects a person draws in flight are shorter and wider than those they draw on the ground.
3. The shortened perception of vertical lines and widened perception of horizontal lines play out in the objects subjects are asked to evaluate. For example, in space, they perceive a reduction in the *T* illusion which on Earth results in a tendency to overestimate the length of the vertical line relative to the horizontal line. The illusion is reduced in space.
4. Finally, in space, the subjects underestimate distances, particularly for distances greater than two meters.

Toward the end of the 3D Space tenure on the station, NPR did a story,[18] saying that even though no more shuttle flights to the ISS were scheduled, the "space station's best days are still ahead." Science correspondent Joe Palca introduced our work: "One thing the space station is good for is understanding how humans function in space. That's what Corinna Lathan is interested in. She's an engineer, neuroscientist, and entrepreneur who has been studying how weightlessness changes the way we perceive the world around us."

The interview gave me the chance to share my experience with robot operations and argue for the importance of understanding human performance in space. I talked about some of the implications

---

18 Joe Palca, "NASA: Space Station's Best Days Are Still Ahead," *All Things Considered*, NPR, July 18, 2011, https://www.npr.org/2011/07/18/138478225/nasa-space-station-s-best-days-are-still-ahead.

of our results. "If you're trying to operate a robot and you have to estimate a distance, you have to use visual cues to guide yourself. If you move your head and get a different input than you're expecting, it could lead to critical mistakes."

We need to understand how the human body works if we want to optimize human–technology interaction. And space is no exception. Our performance changes and we need to design for those new abilities. This will be important as we plan long-duration spaceflight to get us to Mars.

ONCE A TREKKIE...

As I was writing this chapter, I paused as William Shatner, aka Captain Kirk, rocketed to space in Blue Origins' *New Shepard*. The flight was short and he spent only ten minutes in space, but when he exited the capsule, he had tears in his eyes. He said it was "the most profound experience I can imagine." He, like others who have gone to space, spoke about the Earth's vulnerability and the importance of preserving our environment.

I deeply believe that international collaboration will help create a future that we all want to see. And what I learned through 3D Space and my work with scientists and astronauts across the world was that I am uniquely placed to advocate for collaboration as an inventor, an entrepreneur, and a citizen of Earth. True, my fifth-grade prediction hasn't yet come to pass, but if Captain Kirk got a seat on the *New Shepherd*, I'm still not giving up on Mars!

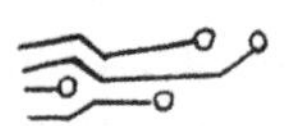

INSPIRATION AND REFLECTION #4
Why Explore Space?

As I worked my way through grad school, one of my jobs was as the night staff person at a men's homeless shelter in Harvard Square in Cambridge, Massachusetts. I will never forget the many nights standing outside with the shelter's clients, pointing out the space shuttle as it passed overhead, docked to the space station. The men enjoyed these moments as much as I did, all of us gazing up at that light in the sky that somehow held people like us inside. Maybe they were transfixed because they had also dreamed of going to space, or maybe for a deeper reason I only came to understand years later.

On the podcast *Mission Interplanetary*, co-hosted by former astronaut Cady Coleman and Professor Andrew Maynard from Arizona State University, Ellen Stofan was asked, "Why do you want to send humans to Mars when you can send robots?" As former director of the Air and Space Museum and now Under Secretary for Science and Research at the Smithsonian Institution, Ellen has led the charge to inspire us with the promise of space exploration. She answered, "One of the reasons is that people can come back and tell stories. They can tell you what they felt. They can tell you what it sounded like."

I agree with Ellen. Stories of space travel universally inspire us as humans to invent the future more than anything else I've encountered. I've spoken to classrooms of students from many backgrounds, all fascinated when I talk about research in space. Space inspires these kids to pursue careers in STEM as well as inspires nations to collaborate on the ISS and the moonshot goal of going to Mars.

Speaking of Mars, another personal inspiration is MIT Professor and Director of the MIT Media Lab, Dr. Dava Newman. Dava and I were graduate students together, and then she became a professor and my master's thesis advisor in aeronautics and astronautics. Dava has designed a spacesuit for Mars exploration, a stunningly beautiful BioSuit that conforms to the body and is constructed to maximize mobility.[19] In addition to adding the *A* and

---

19 Joe Flaherty, "This Spacesuit for Exploring Mars Is a Form-

The Mars BioSuit. (Photo courtesy of Dava Newman.)

D to STEAM'D (Science, Technology, Engineering, Art, Mathematics, and Design), the innovations realized in the BioSuit design are also dual use and can help people with mobility disabilities on Earth.[20]

Fitting Math Problem," Wired, January 7, 2014, https://www.wired.com/2014/01/how-a-textbook-from-1882-will-help-nasa-go-to-mars/.

20 Photo Credits: Professor Dava Newman, MIT: Inventor, Science and Engineering; Guillermo Trotti, A.I.A., Trotti and Associates, Inc. (Cambridge, MA): Design; Dainese (Vicenza, Italy): Fabrication.

Although Dava's BioSuit finally rethinks the bulky 1970s spacesuit design that still exists today, it's amazing to realize that the dream of living and working in space has been a reality for over five decades now. Prior to the Space Shuttle program, NASA launched and operated a research station called Skylab[21] for twenty-four weeks in 1973–1974, before its orbit decayed and it disintegrated in the atmosphere. The Russians operated the Mir space station from 1986 to 2000. And in 1998, the United States, in partnership with Europe, Russia, Canada, and Japan, began construction of the International Space Station, resulting in a permanent presence in space since Expedition 1 launched in the fall of 2000.[22]

The ISS was a technical feat accomplished through international collaboration and is an ultimate example of technology enabling human capability, providing life support in a hostile environment, facilitating research, and promoting international collaboration. Talk about inventing the future! Hundreds of people have now lived and worked in space. They come back and tell their stories, inspiring hundreds or thousands more to dream about going to space—including me and the men at the homeless shelter in Harvard Square.

The current deputy director of NASA is Colonel Pam Melroy, one of only two women to have ever been commander of a space shuttle crew. Pam and I met when we randomly sat next to each other at a brunch in Texas many years ago. We talked the whole time about our favorite science fiction authors—Anne McCaffrey and David Weber, among others. After the brunch, one of her

---

21 "Skylab," Wikipedia, last modified March 3, 2022, https://en.wikipedia.org/wiki/Skylab.

22 "International Space Station legal framework" European Space Agency, accessed March 18 2022, https://www.esa.int/Science_Exploration/Human_and_Robotic_Exploration/International_Space_Station/International_Space_Station_legal_framework.

colleagues asked her, "How long have you known Cori?" Of course she replied, "Oh, we just met!" But we were drawn together by our shared love of stories that depicted where spaceflight could take us.

Pam knows from experience in space how critical exploration is—not just for the space program but for each of us. At her induction into the Astronaut Hall of Fame in 2021, she said, "Now we are building a program to achieve a series of objectives that will provide the blueprint to how we maintain a human presence in deep space as we explore the solar system, and then the universe. Those of us who had the incredible good fortune to go to space know that all of humanity is the crew of spaceship Earth, and it is our duty to care for each other and our home planet."[23]

23 NASA/Kim Shiflett, "NASA Deputy Administrator among US Astronaut Hall of Fame Inductees," November 13, 2021, https://www.nasa.gov/press-release/nasa-deputy-administrator-among-us-astronaut-hall-of-fame-inductees.

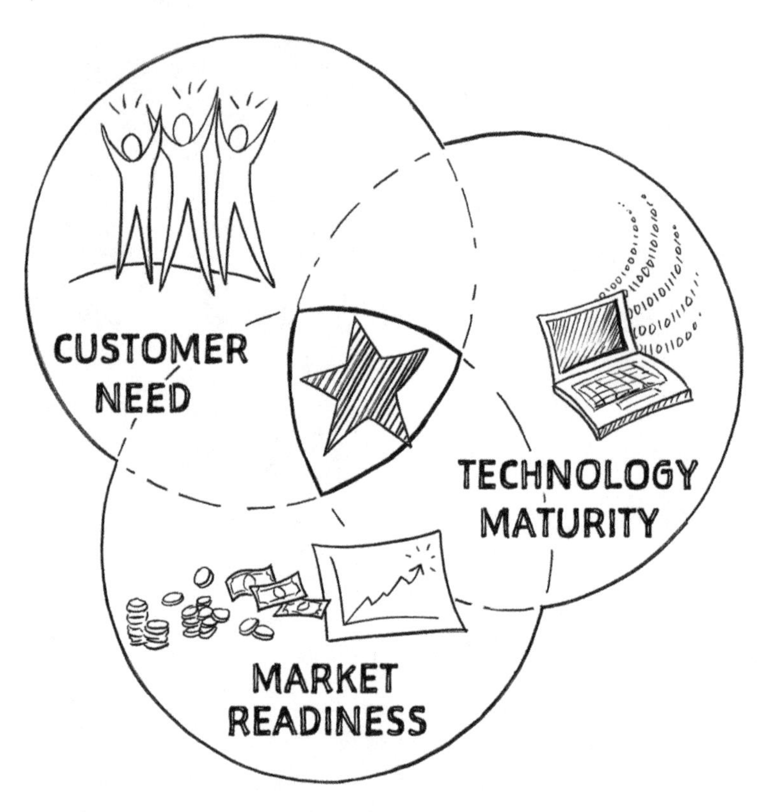

IF your product fails in the market,
THEN your efforts will inspire the next attempt.

CHAPTER 5

Tech and Timing

THE STORY OF VISUNIT

REMEMBER GOOGLE GLASS? Google invested millions of dollars and came to market in 2014 with the first generation of wearable augmented reality (AR) technology. You could wear their cool headsets with or without glasses and use them to read and send email all while going about other tasks.

Even before it was available to the public, Glass created a huge amount of buzz. In 2013, everyone wanted to get in on beta testing. Almost every keynote speaker I saw that year would come out wearing a Google Glass headset—and then invariably admit the battery had run out weeks ago and they were just wearing it to look cool. I have to admit, I did the same.

Maybe that's because it made me nostalgic for a product we had designed about ten years earlier, in 2005. Our Visually Integrated Sensor Unit, or VISUnit, was an early version of an AR headset. Everyone agrees that Google Glass failed when it came to market, but to me, it's a symbol of hope. Just the fact that it got to market makes me even prouder of VISUnit, even though our journey was quite a bit different. Glass may have inspired the next generation of AR developers commercially, but our work was likely the first to demonstrate the feasibility of augmented vision systems in the military. Let me back up...

Robots and the Military

By the early 2000s, we had made a name for ourselves in the DoD research and development world, thanks to the AcceleGlove and a few other projects. We had built connections at DARPA and with various defense contractors and were ready to help the US military move into the future. And the future, according to the DoD, included robots, or in military-speak, UGVs—unmanned ground vehicles.

One of the funders of the Tactical Mobile Robots program had been the DoD's Joint Robotics Program (JRP). They established the program in 1990 to provide the military with "leap-ahead" capability for the twenty-first century through investment in ground robotics. The JRP envisioned teams of small UGVs for reconnaissance, surveillance, logistics, transport, and even as weapons platforms. They defined a UGV as a mobility platform with sensors, computers, software, communications, and power.

The National Research Council (the research arm of the US National Academies of Sciences, Engineering, and Medicine) released a report in 2002 saying that if UGVs were "developed to their full potential, their use would reduce casualties and vastly increase combat effectiveness."[24] The report recommended devel-

24 National Research Council, *Technology Development for Army Unmanned Ground Vehicles* (Washington, DC: National Academy Press, 2002), https://www.nap.edu/catalog/10592/technology-development-for-army-unmanned-ground-vehicles.

opment of both autonomous and human operator capabilities. The JRP was now conceptualizing how to put together teams of humans and robots and what type of technology could serve as an interface between them.

At AnthroTronix, it was clear UGVs would be increasingly important to the DoD's effort to decrease casualties and loss of life and serve as important tools for soldiers. If we could leverage our interface expertise and our defense connections, we could supply multiple contractors with the means to connect humans with the unmanned vehicles they needed to control. All we had to do was build the perfect device. The investment would pay off in dividends if we could become a robot control supplier to multiple defense contractors.

The Problem We Were Trying to Solve

Traditional robot controllers for the military only allow a single operator to control a single UGV, and the operator needs special training to control it. For this reason, UGVs are still used for a limited number of tasks.

In the early 2000s, with wars in Iraq and Afghanistan, troop security was top of mind. Securing bases of military operations, both big and small, was a key job of military personnel, so the use case for UGVs was clear. A human–robot team could investigate threats more safely.

We received a contract through a DoD mentorship program that partnered large companies with small businesses to encourage innovation. Lockheed-Martin's Advanced Technology Laboratory received funding over three years to mentor us to develop a next-generation controller for the teleoperation of multiple UGVs.

The idea was that a soldier of the future would need to monitor and potentially control multiple UGVs while in the field. We knew the soldier would need something that gave them information about their immediate surroundings as well as the remote location and perspective of the robots under their supervision. Through the Multi-purpose Autonomous Teaming Control of Heterogeneous Robots (MATCH) project, we proposed to build one.

Our Solution: The VISUnit

Our vision was to create a single, lightweight, unmanned vehicle control system that would allow warfighters to command a team of robots, which would be autonomous most of the time, yet could be controlled directly when necessary, and that would require as little training as possible to operate.

We called it a Visually Integrated Sensor Unit, or VISUnit. It could be controlled by either motion or speech and could also run autonomously, using sophisticated software to track the location of each human and robotic team member and to prioritize tasks.

The initial VISUnit concept shows a soldier monitoring three different robots through a monocular version of the VISUnit. Holding it up to the eye activates it, like how your phone display comes on when you pick it up. To track what a remote robot is seeing, you move your head; your view pans and tilts correspondingly. To interact with the robot, you use a thumb joystick and finger buttons.

CONCEPT OF OPERATION

As a special operations marine, my co-founder Jack Vice was trained in counterterrorist and unconventional warfare methods. He was perfect for developing the concept of operations, which essentially describes how a system will work from the point of view of the user. Jack had seen the need for these tools firsthand, running security and reconnaissance missions during Operation Desert Storm in Iraq.

He was all too familiar with the ongoing threat of terrorist attacks on military installations at home and abroad, including IEDs, rocket-propelled grenades, and random insurgent attacks. Patrols

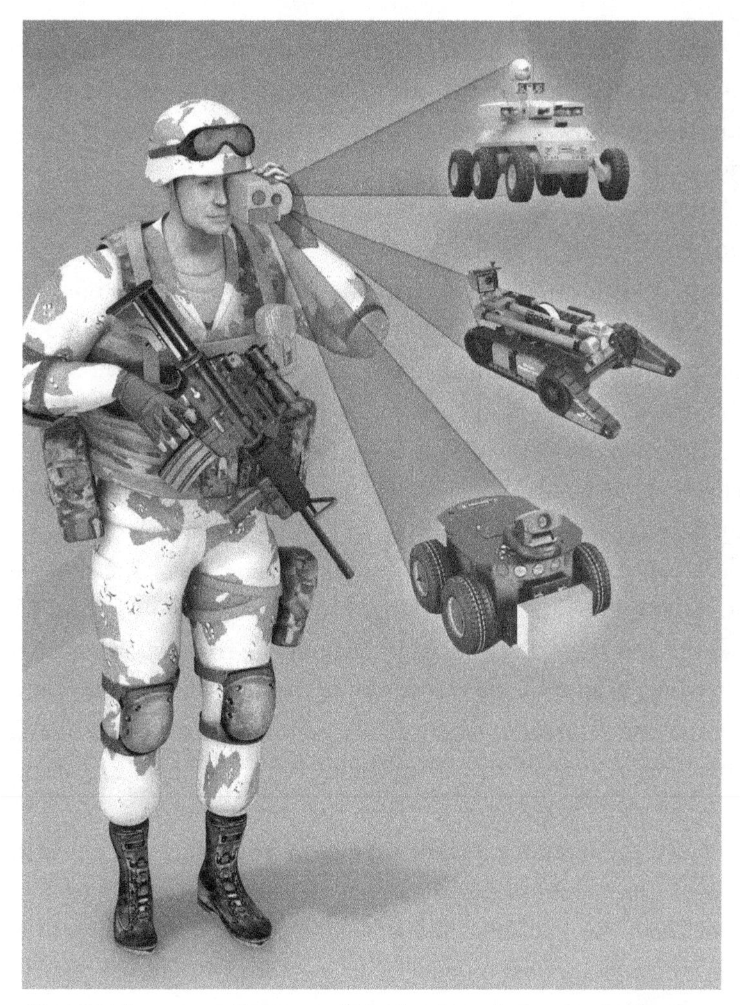

The VISUnit concept. (Courtesy of Lockheed Martin Corporation.)

were sent out daily to assess the security of the base and identify potential threats. Jack's experience helped us imagine what these "force protection patrols" could look like with a VISUnit.

The following scenario is an excerpt from a proposed VISUnit concept of operation to show not only how the equipment works but also what the user would experience.

During the course of a force protection patrol, a team discovers an abandoned truck on the side of the road. Suspicious, the operator decides he should mark this truck for further investigation by one of the team's robots.

The operator lifts the VISUnit up to his eye, triggering it to wake up.

The VISUnit starts up in Local Mode and presents the operator with several options. From the list of actions available, the operator uses the context-sensitive soft buttons to quickly switch into Enhanced Local Mode, activating the infrared camera.

The operator immediately sees that the abandoned vehicle's engine is still emitting heat, indicating that it has only been present for a short period of time.

Because of this, the operator now switches back to Local Mode to enter a Potential Threat Designation. In this mode, he can target the area to be investigated by aligning the crosshairs and either pressing the appropriate soft button or speaking a command. Once the potential threat is targeted, the operator then assigns one of three priority levels—high, medium, or low—which are distinguished with the colors red, yellow, and green.

Next, the operator assigns a robot to investigate the potential threat. The robot assignment screen presents a variety of information about the available robots, including their distance from the targeted area, current number of active tasks, and sensing capabilities. Based on those parameters, the system recommends a robot.

The operator can choose to accept, reject, or modify the recommended selection. Once a robot is chosen, an overlay, in the form of a yellow diamond, appears. The operator can continue with his own mission while the robot carries out the task.

Once the robot reaches its location, the operator is notified by both an audible tone and a vibration. The operator once again activates the VISUnit by raising it to his eye. The system now offers the option to switch to Remote Mode. In this mode, the operator can see through the robot's viewfinder and also has the option to drive the robot's motors.

After remotely inspecting the area, the operator determines whether the potential threat is actually an area of concern—in which case, he marks it as a threat. Once this occurs, the coordinates are displayed for reference and the overlay transitions from yellow to red, identifying the area as being dangerous. And all without a human being entering the area of potential threat or having to manually control the robot every step of the way.

End of the Road

We demonstrated our VISUnit technology at AUVSI 2007, the annual expo of the Association for Unmanned Vehicle Systems International, in Washington, DC. At that point, no one had yet heard of augmented reality, so the reactions were...interesting. People would say things like, "It's a combination of binoculars and night vision goggles, right?" A few people actually understood what the VISUnit was fully capable of and had the much appreciated "Oh wow!" reaction. "It's like something out of a sci-fi movie!"

Despite the level of interest and enthusiasm, we were in a tough spot. Without significant financial investment, further development was beyond our reach. Our DoD contract didn't cover production and the UGV market was too small to support outside investment. We and our mentors at Lockheed-Martin believed in the potential growth of the UGV market and thought it was on the verge of taking off. We were wrong.

In fact, two decades after the DARPA TMR program, the UGV market is still not much larger. A 2021 article reported that "UGVs will certainly be a welcome addition to supporting soldiers on the battlefield, but the need for complex systems is still being evaluated."[25] In the same story, an army general describes UGVs as being "relatively new territory."

There are merely thousands of military UGVs deployed, instead of the tens of thousands envisioned by the 2002 National Research Council report. The vision of supporting soldiers in the field with this remote technology has been realized only in limited cases. For instance, UGVs are used in explosive ordnance disposal (EOD)—that is, to defuse bombs. But EOD systems follow the traditional model of one highly trained operator per one specialized robot.

So why didn't UGVs take off? For a technology or product to take off, the need, market, and technology all need to align at the right time.

Meanwhile, in Autonomous Navigation

No invention exists in a vacuum. In 2007, the same year we presented the VISUnit at AUVSI, DARPA held its urban challenge across the country at an abandoned air force base in California. Teams were challenged to build an autonomous vehicle capable of driving in traffic and performing complex maneuvers such as merging, passing, parking, and negotiating intersections.

Six robotic vehicles completed the challenge, proving that driverless cars could become a reality—even on city streets. This launched the commercial field of self-driving cars that Tesla, Google, Uber, and Apple are all investing in so heavily. Early-use cases like Waymo in Phoenix and Cruise in San Francisco have proven the concept is sound and the market exists, although there are still plenty of kinks to work out.

---

25 Stephen W. Miller, "Battlefield UGVs Make Steady Progress," *Armada International*, last modified February 2, 2021, https://www.armadainternational.com/2021/02/battlefield-ugvs-make-steady-progress/.

Military conditions present unique challenges to autonomous navigation. Compare the relative order of a traffic grid to an unstructured environment like a forest or mountainous terrain, or the chaos and rubble of a war zone. For these conditions, there's a better solution: take the autonomous vehicle off the ground and into the air. So while the autonomous vehicle community turned its attention to the commercial world of self-driving cars, the military went in another direction: drones instead of ground robots.

Since the start of the Gulf War, the military had been using a few expensive unmanned aerial vehicles (UAVs), or drones, such as the Predator drone that was used in the hunt for Osama bin Laden in 2000. But the technology advanced quickly, enabling a sharp increase in use by 2010. So in the few years since the DoD's Tactical Mobile Robots program had launched the UGV field, the market had shifted. Today, the military manages battlefield surveillance with thousands of lightweight drones—many can even be operated using a smartphone.

Technology Is a Bridge

I've talked about how technology should enable human capability. It should take us from what we *can* do to what we *want* to do. In an ideal world, technology bridges capabilities.

Technology is also a bridge in time.

An invention doesn't always receive recognition or gain an immediate user base. But it captures and documents a moment that gives rise to the next generation of creators.

The world of technology doesn't always think this way. People argue over who invented something first. Patents are a way to put a stake in the ground, but that's all they are. Just because you patent something doesn't mean it will be a viable product. Although timing is probably one of the most critical factors in the success of technology, it is impossible to predict or control.

What's interesting with Google Glass is that they probably suspected the timing was wrong, but they tried anyway.

Bringing a tech product to market successfully requires a sweet spot in the Venn diagram overlap of customer need, technology maturity, and market readiness.

If you think of the inventions discussed in previous chapters, you can see why market success is almost always a long shot. In the case of CosmoBot, despite the high need for innovative, interactive educational technologies for kids with disabilities, the technology was not mature. Consumers were not ready to pay the price. So we switched gears and developed Cosmo's Learning System to hit the sweet spot of need, tech maturity, and market.

But that Venn diagram, pictured at the chapter beginning, is not static. It's constantly changing. In the case of CLS, the technology and therefore the market changed even though the need did not. As the iPad and other tablet computers became the norm, the tablets drove and changed the need—even though their use was not initially backed by science.

The AcceleGlove flaunted a similar model to Google Glass. We put innovative technology out into the world with some demos and a developer's kit and let the market create the need. As with Google Glass, the product didn't make it very far. But both became catalysts for change, albeit on different scales. The AcceleGlove bridged our ideas for gestural interfaces to applications developed by those who purchased the kit or saw a demo. Google Glass gave hundreds of innovators a chance to experience the potential of augmented reality.

Both inventions set the stage for killer apps and compelling use cases. In the case of gestural interfaces, Nintendo and Microsoft later made gestural interfaces the norm for video gaming, with Wii and Kinect. Just as Oculus and Hololens are now making similar inroads for VR/AR. And technology does come back around. Microsoft recently won a $22 billion contract from the military to develop the Integrated Visual Augmentation System, which is an AR system for soldiers' decision making. Sound familiar?[26]

On the consumer side, the fact that Google Glass existed was enough. It inspired people to think beyond traditional ways of moving about our environment. Instead of looking down at a device, we could look out at the world and have an overlay of digital information. It was one of the first steps in physical–digital convergence that we are seeing today throughout many industries. And as someone who has created many inventions that never left the lab or were too early for the market, I love the fact that Glass came out at all!

Facebook, Ray-Ban, and Snap have released smart glasses, which contain embedded cameras. The companies are betting that focusing on the form factor of cool shades and the convenience of being able to take a photo quickly and easily, while not cutting edge, may in fact be the killer app that finally breaks the consumer market open.

Time will tell.

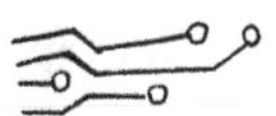

INSPIRATION AND REFLECTION #5
Digital and Physical Convergence in the Metaverse

Imagine you have moved to a new city and bought a condo advertised as "metaverse enabled." Upon closing, along with the physical keys to your condo, you receive a unique cryptographic

26 Alex Kipman, "Army Moves Microsoft HoloLens-based Headset from Prototyping to Production Phase," *Official Microsoft Blog*, March 31, 2021, https://blogs.microsoft.com/blog/2021/03/31/army-moves-microsoft-hololens-based-headset-from-prototyping-to-production-phase/.

key to the community. You move into your neighborhood both physically and digitally. With the crypto key, you link your own personal metaverse profile to the condo. This merges your personal metaverse, including a digital catalog of all the objects in your home, to a digital map of the new space integrating all the sensors and devices that control the objects through the metaverse. As you move from your bedroom to the kitchen, the lights switch on, and the room temperature adjusts. In the kitchen, you tell the stove to turn on and a graphical display appears in front of you as you turn your gaze toward the refrigerator. You reach out your hand and select the egg icon, which reveals that you have three eggs left. You swipe up with your hand and the order goes out for more eggs.

As you prepare breakfast, you gesture again, and in another moment your mother's face appears in front of you. "Mom, do you want to join me for breakfast?" "Sure," she responds as she turns down your stove so your eggs don't burn! You gather your eggs and bacon and sit down at the table, which has moved out from a recess in the wall in anticipation of your approach. You sit down, your mother's hologram appears across from you with her breakfast, and the two of you begin to discuss your day.

Later that evening as you walk around the neighborhood guided by your AR display, you're prompted with information tagged to houses throughout the neighborhood. One that catches your eye says, "Book club meeting at 6:30 p.m. tomorrow at my house—digital only this week, meet in the community metaverse lounge. New members welcome!" You'll fit right in!

Ready to move into this condo? What I'm really describing is a form of spatial computing—technology that seamlessly integrates the physical and virtual world. I wrote about spatial computing in a *Scientific American* article in 2020, and it's one of the building blocks of the metaverse, the vast network of physical and virtual networks primed to redefine the way we live and work in the not-so-distant future.[27]

You may have glimpsed the metaverse already if you've watched the movie *Ready Player One* or played *Minecraft* or *Fortnite.* Massively multiplayer online games like *Fortnite* have been around for over two decades, as have online worlds like Second Life. Around the time I started AnthroTronix in 1999, Jacki Morie, a virtual reality pioneer, helped establish the University of Southern California's Institute for Creative Technologies, an army-funded research lab finding connections between entertainment and military needs. She started the Coming Home project in 2009, creating a center in Second Life where veterans and soldiers could find stress relief and rehabilitation activities. Kindred spirits in the quest to use technology to create meaningful experiences that enable ability and enrich people's lives, we began a collaboration. We connected our digital glove technology to Second Life, enabling veterans using wheelchairs to rock-climb, and our teams jointly invented a "scent necklace" that released computer-controlled scents to make virtual reality worlds more immersive.

We don't really know the potential of the metaverse, or whether it will be completely virtual or a mix of physical and digital. As we leverage the internet, advances in virtual and augmented reality technologies, and spatial computing, there will likely be multiple metaverses, including a potential "Medi-verse" that focuses on healthcare delivery. But whether you enter a consumer-oriented gaming metaverse or an immersive work one, one thing that seems

---

27 Corinna Lathan and Geoffrey Ling, "Spatial Computing Could Be the Next Big Thing," *Scientific American,* November 10, 2020, https://www.scientificamerican.com/article/spatial-computing-could-be-the-next-big-thing/.

unlikely is us continuing to interact with the digital world through desktop computers, smartphones, and bulky headsets. More likely, we will interact with the digital world through voice recognition, eye tracking, and natural gestures like in the scenario above.

My vision of the metaverse is that we will be able to physically go about our daily activities while interacting in real time with friends and colleagues around the world. I want to go into a business meeting and sit around a table with colleagues from different countries without fighting with video conferencing displays. I want to go to the gym and work out with my friends in Paris and San Francisco.

There is a much-used quote attributed to many, including Abraham Lincoln: "The best way to predict the future is to create it." We have the tools to create the metaverse. What we do with them will be up to us.

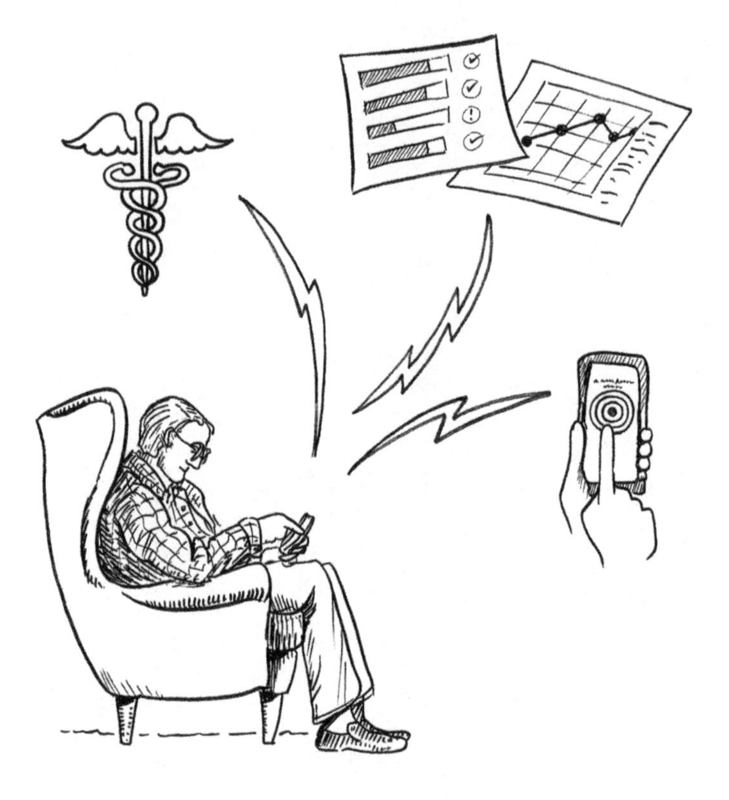

IF you fix a flaw in the system,
THEN you can disrupt an entire industry.

CHAPTER 6

Games for Brains

THE STORY OF DANA

FOR AS LONG as I can remember, my father loved acting. Into his sixties and early seventies, he was quite active in the theater. He played Tartuffe in Molière's *Tartuffe*, Nick Bottom in Shakespeare's *A Midsummer Night's Dream*, and the Old Man in Steve Martin's *Picasso at the Lapin Agile*. When he won the role of Scrooge in a local production of Dickens's *A Christmas Carol*, I was so excited for him that I bought tickets way before opening night. But he was having trouble remembering his lines. Eventually, the director had to let him go.

To find out what was going on, my mom and dad went to his primary care physician, who referred him to a neurologist. After waiting a month for that appointment, the neurologist told Dad to see a neuropsychologist, who was booked another three months out. When that appointment arrived, the neuropsychologist gave him a variety of cognitive tests, including written, verbal, and computer based. After another month, the neurologist called us back in and told my father, "You have mild cognitive impairment."

"No shit, Sherlock," I thought. "That's why we went to see his doctor six months ago." The neurologist then discussed my father's other health issues with us, which included cardiovascular disease, sleep apnea, type 2 diabetes, and by that point, depression. I then had an inspiration. "Dad, when was the last time you used your CPAP machine?" He admitted sheepishly, "I don't use it. I don't like it."

"Dad, if your brain doesn't get oxygen at night, it's not going to work during the day." Problem solved. Well, not totally, but it was a start. The real problem was that everything my father was going through—from mental health to chronic illnesses and aging—affects brain health. As a neuroscientist, I think about these issues a lot, but this really hit home, especially in light of a project AnthroTronix was tackling with the military. Let me back up...

Measuring Brain Health: The Problem We Were Trying to Solve

In the early 2000s, brain health took center stage in the media. Three major factors brought awareness to the issue: concussion, suicide, and Alzheimer's disease. I'll talk more about Alzheimer's later. For now, suffice it to say that almost 6 million Americans live with the disease, and more than 200,000 die each year of Alzheimer's and related dementias.[28]

With the release by the NFL Players Association of the first study on the long-term impact of head injuries, concussions

---

28 "Alzheimer's Disease and Related Dementias," Alzheimer's Disease and Healthy Aging, Centers for Disease Control and Prevention, last reviewed October 26, 2020, https://www.cdc.gov/aging/aginginfo/alzheimers.htm.

became a big topic in both professional and amateur sports.[29] Around the country, parents started questioning whether they should even let their kids play football and soccer. My own daughters participated in sports, and I worried particularly about my fourth grader, a gymnast. Ironically, she got a serious concussion not from doing a backflip but from falling off a piece of playground equipment.

It took her three months to recover. During that time, I learned more about concussion treatment than I ever wanted to know. For someone who likes to analyze and solve problems, I found it particularly distressing that there is *no* single diagnostic for a concussion. No blood test or imaging technique. Instead, the diagnosis is a subjective determination based on multiple factors:

- Circumstantial (i.e., a recent blow to the head).
- Symptoms (e.g., a headache).
- Impaired balance (as measured by dizziness or stumbling).
- Impaired cognition (as measured by simple verbal questioning).

You'll note the last three criteria are not very specific to concussion. If the first criterion isn't met (that is, if you didn't recently hit your head), a doctor wouldn't necessarily diagnose concussion as the cause for your symptoms.

The media was also paying attention to the rising suicide rate. Statistics from the CDC website show that since 1999, suicide rates in the United States have increased 33 percent. Suicide has become the second leading cause of death for people ages ten to thirty-four,

---

29 American Academy of Neurology, "Concussions May Spell Later Trouble for Football Players," ScienceDaily, accessed February 28, 2022, www.sciencedaily.com/releases/2000/05/000505064356.htm.

the fourth leading cause among people ages thirty-five to forty-four, and the tenth leading cause of death overall.[30]

Both crises affected and impacted the military in historic ways.

BRAIN HEALTH AND THE MILITARY

After the events of 9/11, serious forms of concussion known as traumatic brain injury (TBI) and acquired brain injury (ABI) in the military increased threefold. Most TBIs and ABIs were caused by exposure to improvised explosive devices (IEDs). The military was also facing an increase in suicides, which nearly doubled from 2001 to 2011, to about 300 per year.

There was more than correlation between these increases. In many cases, a TBI or ABI would linger for years, the symptoms going undiagnosed or untreated, and triggering or exacerbating other mental health issues until they became unbearable. For example, Kim Agar was a twenty-five-year-old Texan who joined the military in 2006 and served in Iraq. Kim sustained a serious concussion when her truck was hit by an IED in 2007. She died by suicide in 2011.[31]

That same year, the DoD established the Defense Suicide Prevention Office to try and address the complex problem that shortened the lives of so many soldiers like Kim who had served their country.

Prior to 2011, as part of the 2008 National Defense Authorization Act,[32] the DoD had stated that the "Secretary of Defense shall

---

30 "Facts about Suicide," Suicide Prevention, Centers for Desease Control and Prevention, last reviewed February 24, 2022, https://www.cdc.gov/suicide/facts/index.html.

31 Sarah Mervosh, "A Year Later, Texas Soldier's Suicide Still Haunts Her Mother in Bedford," *Dallas Morning News*, July, 03 2013, https://msrc.fsu.edu/news/year-later-texas-soldier%E2%80%99s-suicide-still-haunts-her-mother-bedford.

32 US Congress, House, *National Defense Authorization Act for Fiscal*

develop and implement a comprehensive policy on consistent neurological cognitive assessments of members of the Armed Forces before and after deployment." In other words, before a soldier could be deployed, they would have to complete a set of cognitive tests to serve as a baseline. They would repeat the same testing after deployment if they were exposed to a blast and a TBI was suspected.

This congressional act had good intentions. However, the guidelines assumed that TBIs were the only factors that affect brain health. But when our soldiers are deployed, they face *many* impactful conditions. They regularly cope with situations that at a minimum cause combat fatigue and stress. A soldier can also experience insomnia, depression, trauma, and yes, blast exposure. *All* these things can change cognitive function.

Medical professionals in the military realized this, so despite the congressional mandate to administer these tests, the results were largely ignored.

By 2010, at the peak of the concussion crisis and the suicide crisis, it was clear that a new approach to brain health was needed. The Navy Bureau of Medicine contracted AnthroTronix to develop a tool to measure the impact of deployment as a whole—not just blast exposure—on cognition.

Our Solution: Taking a Step Back

Once we received the contract to track brain health, we worked with an advisory board made up of members from the army, navy, air force, marines, and coast guard, as well as civilian academics in the field.

We examined how experts currently measured brain health—like the neurologist who met my dad, gave him some tests, and

---

Year 2008, HR 4986, 110th Congress, January 28, 2008, https://www.govinfo.gov/content/pkg/PLAW-110publ181/html/PLAW-110publ181.htm.

made a determination. No matter how gifted the neurologist may be, there was no way for him to know whether my dad's cognition had declined, improved, or stayed the same over the past months or even years. Yet I'd expected that doctor to have all the answers.

We proposed a new approach. Rather than boil the ocean trying to characterize every aspect of cognition, we said, "Let's track brain health as if it's a vital sign." To do this, we needed a tracking tool that was as easy to use as, say, a thermometer or blood pressure cuff. Before you assume this is impossible, think about the way you actually use a thermometer. The readout tells you whether your temperature is high or low, which helps you to form a picture of your overall health. You would never automatically attribute a high number to a particular illness, though a fever is certainly a sign that something may be severely wrong.

With a blood pressure cuff, we would never check your pressure once—or even twice—to decide whether you have hypertension. We would need to track it over time so we have a fuller picture. Having that baseline also helps us determine if a change occurs that is significant for you. For example, I have consistently low blood pressure, so what's high for me may be normal for you. And if we didn't have that baseline, a doctor might consider my blood pressure to be fine when in fact it's elevated for me.

The answer our tool would provide was *not* whether the soldier had a concussion or depression, but whether there had been a *change* in their brain health. The appropriate qualified medical personnel would then be alerted to make an actual diagnosis.

Just as a thermometer reading doesn't tell you what's wrong with you, we wouldn't expect this tool to diagnose a soldier. But like a thermometer, it would quickly and reliably assess meaningful changes in a soldier's medical status.

Other key aspects of our approach were that the tool must work on any mobile platform *and* be self-administered. Our concept of operations was that a busy medic in the field could give a soldier a handheld device, walk away to do whatever else they needed to do, and come back five minutes later for the answer.

So what did the tool actually *do*?

Building DANA, the Brain Vital

Some of you may have tried "brain games." These are simple activities that ask you to respond quickly to simple questions, like picking shapes that match or indicating which object is different from the others. They measure both reaction time (how quickly you respond to each question) and accuracy (how many you get right). These "games" are actually modeled on brain assessments that have been around for decades, originally as paper-and-pencil tests and then translated to computer for both assessments and their gamified versions.

Working with our advisory board and combing through the published research, we looked at all the brain assessments we could find and the science behind them. We wanted to find the most sensitive and reliable tests not to improve the brain but to assess changes in its health.

Our team found three tests that met the requirements. All had been used in research for over twenty years, so they were time tested and refined. One measured simple reaction time: when an object appears in the middle of the screen, touch it as quickly as you can. The second measured choice reaction time: when an object appears in the middle of the screen, choose the object below that matches it. The third, a go/no-go game, measured impulsiveness: you are shown an object to target and must touch it as quickly as you can when it appears on-screen (go) and do nothing when a nontarget object appears (no-go).

Simple, right? They are supposed to be. You shouldn't need a college education to do well; you just need sustained attention. Our goal was to measure focus or attention and how quickly someone could respond accurately, called processing speed. Attention and processing speed are building blocks for the harder stuff our brains have to do, like remembering instructions or planning steps to complete a task.

Using those three games, we developed an app-based cognitive assessment tool, a "brain vital sign" for military use. We called the app DANA.

GETTING DANA APPROVED

The FDA regulates medical devices, so getting our device "cleared" by the FDA was required if we wanted to distribute DANA to any medical institution, military or otherwise. But in 2013, mobile medical apps were so new that the FDA did not yet have guidelines for applying. They were essentially demanding compliance but couldn't yet tell us exactly what we needed to comply with. As one of the very first apps to go through clearance, we had the obligation and the opportunity to work with the FDA on the process. Over two years, we worked with them to establish what it meant for an app to essentially be a medical diagnostic support tool.

We positioned DANA as a platform that could run reaction time–based tests and provide that data to a clinician to aid in their diagnosis.

We wanted to be able to market our app on any mobile device, so we needed clearance for a device-agnostic platform, not for one particular piece of hardware. We also made a case for allowing it to be remotely and self-administered. This would later become critical, enabling us to partner with hospitals for home use during COVID-19.

In October 2014, we delivered an FDA-cleared cognitive assessment tool, a "brain vital sign," to the military. Although we'd originally named it DANA as an acronym for Defense Automated Neurobehavioral Assessment, we now leaned into a more dual-

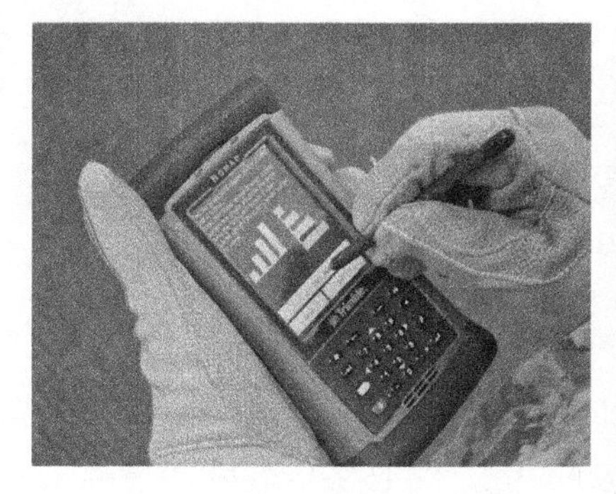

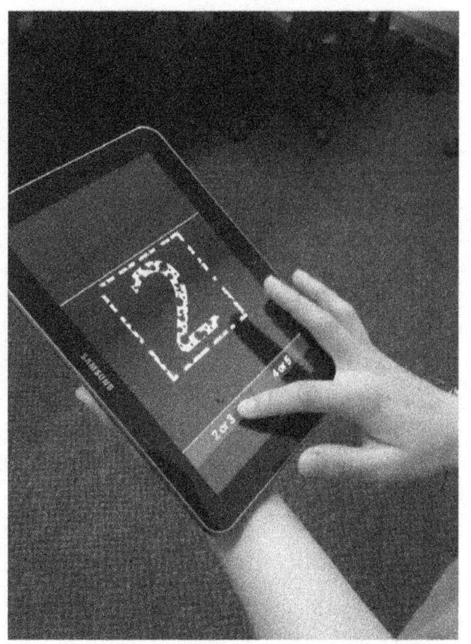

The DANA tool. Military and commercial versions. (Photo courtesy of AnthroTronix.)

purpose meaning of the word: the D in DANA also stands for "digital," and as an added bonus, "dana" means "wisdom" in Persian.

DANA beyond the Military

The military invested millions of dollars in the development and validation of DANA and gained the right to use it whenever and however they chose. My company has no control over how or even whether they use DANA. However, we did retain the right to commercialize DANA for nonmilitary applications, and that's what we did next.

Our intensive validation process had involved testing thousands of active-duty soldiers. We had documented effects of insomnia, depression, and PTSD on cognition—all things that are prevalent in nonmilitary populations as well. Knowing the potential for everyone to benefit from DANA, we couldn't just walk away.

We'd also learned through our work with the military that to get the highest sensitivity for detecting meaningful change in brain health, you had to compare an individual only to themselves. As soon as you averaged the results over a bunch of people, you'd lose sensitivity.[33] Some people in the group might be sleep deprived, have depression, or be taking medication.

DANA FOR ELECTROCONVULSIVE THERAPY

One of our colleagues at Johns Hopkins, Dr. Adam Kaplin, got it! As a neuropsychiatrist who treated patients with depression, he was frustrated that there was no good method to measure changes to a patient's cognition during treatment. He saw the need for a way to do this both bedside for inpatient treatment and in private homes for outpatients.

The standard method, called a mini-mental, is a list of questions a clinician can ask like, "What month is it?" or "Where are we now?" The total possible score is thirty, and anything above twenty-four is considered normal. It is hard to fail a mini-mental unless you are greatly impaired.

We started working with Dr. Kaplin to evaluate DANA as a more sensitive tool. Some of his toughest patients were suffering with major depressive disorder. Medication hadn't worked, and the next treatment option was electroconvulsive therapy (ECT). Often maligned as "shock treatment," ECT is a scary, last-resort therapy that can affect not only attention and processing speed but also your memory. One of my fellow students in grad school had received ECT, and although it did help relieve her depression, heartbreakingly she could no longer remember her own wedding.

One of the reasons ECT is scary is the inability to measure what is happening to your brain during treatment. Dr. Kaplin believed DANA could change that. He ran a study with seventeen patients

---

33 For a great book on this topic, read Todd Rose, *The End of Average (New York: Harper Collins, 2016).*

who had failed to respond to pharmacological intervention and were now trying ECT. He administered DANA a couple of times a week, as well as the mini-mental, during their course of therapy.

There were two major results. The first was that using the mini-mental test, many patients tested as normal throughout their treatment. Remember, they only had to score above twenty-four, and the questions were not challenging. But DANA was not concerned with the patient passing or failing a set number. Its score combines average reaction time and number of correct responses. What's meaningful is whether the score improves, gets worse, or stays the same throughout treatment.

Patients did all three. Some improved their attention and processing speed during treatment, some showed no change, and some got worse. This last category was important because a doctor could potentially look at the dose of ECT and change it so a patient stops declining.

Dr. Kaplin published these results, and we started presenting the "brain vital" approach at meetings and conferences across the country.[34]

DANA for Alzheimer's

At one of these events, a healthcare roundtable held by then–Maryland governor Martin O'Malley, I talked about the importance of tracking brain health throughout our life span just as we track our height and weight.

---

34 Kristen R. Hollinger, Steven R. Woods, Alexis Adams-Clark, So Yung Choi, Caroline L. Franke, Ryoko Susukida, Carol Thompson, Irving M. Reti, and Adam I. Kaplin, "Defense Automated Neurobehavioral Assessment Accurately Measures Cognition in Patients Undergoing Electroconvulsive Therapy for Major Depressive Disorder," *The Journal of ECT* 34, no. 1 (2018): 14–20.

It turned out that George Vradenburg, a former AOL executive, was there with his wife, Trish. George and Trish had founded the organization UsAgainstAlzheimer's after their experience with Trish's mother, who had died of the disease.

According to the CDC, rates of Alzheimer's are predicted to rise from about 6 million today to about 14 million by 2060, so early detection will be increasingly important to managing this and related diseases. George and Trish immediately saw the potential to use DANA for early detection, and we started exploring opportunities.

They introduced me to a wonderful group of Alzheimer's advocates, including journalist Meryl Comer, author of *Slow Dancing with a Stranger*, an account of caring for her husband who'd developed early-onset Alzheimer's, and Stacy Haller, CEO of the BrightFocus Foundation, which supports research on brain and vision-related diseases.

Meryl advocates for caregivers of those with Alzheimer's, who are themselves at high risk for health conditions like depression and Alzheimer's, in part due to the stress of caregiving. We designed a set of studies funded by BrightFocus to study Alzheimer's patients *and* their caregivers at the same time using DANA and other cognitive testing tools.

There were three important findings. First, caregivers as a population did not perform as well on cognitive tests as matched controls, although their results fell within what was considered normal range. Second, they performed better than their loved ones with Alzheimer's—not surprising but still significant for our purposes. Finally, both caregivers and Alzheimer's patients were able to successfully do the cognitive testing from their home.

This work paved the way to funding from the National Institutes of Health to work with Dr. Rhoda Au, who directs cognitive testing for the famous multigenerational Framingham Study and the

Alzheimer's Disease Center at Boston University. As of this writing, we are two years into a study following a cohort of older adults with mild cognitive impairment to see if DANA can detect cognitive changes *before* a dementia diagnosis.

And our work testing the impact of DANA didn't stop with Alzheimer's research. We received an additional NIH grant to work with the Johns Hopkins Medicine's virtual COVID-19 clinic to track recovering patients who might experience diagnosed or undiagnosed "long-haul" symptoms like cognitive impairment. This research is ongoing and will document and hopefully help many people who are struggling with the lingering effects of COVID-19.

DANA B2B

In some ways, DANA was the least technologically advanced project we ever did at AnthroTronix. It is a simple cell phone app, yet the science, regulation, and disruptive applications within healthcare all combined to make it the hardest and possibly most impactful project we'd ever done to that point.

From a business perspective, we realized that marketing the commercial product DANA takes a skillset very different from the one we've developed at AnthroTronix, where we focus on building and testing innovative technology. So we decided to pursue a business-to-business (B2B) licensing strategy.

DANA's superpower is to track and detect small changes in an individual's cognitive function over time. Many companies have recognized this as an important piece of their own market strategy. For example, DANA is part of the ClearEdge Toolkit, a suite of tests and assessments developed with SUNY Upstate Medical Center's Concussion Clinic to measure and track student athletes. The primary application is to evaluate them for return to school and return to play after a concussion.

Another licensee of DANA is Linus Health in Boston. Linus is a platform that integrates multiple neurocognitive monitoring tools for clinicians to conduct bedside assessments.

One more that I'm excited about is MindMaze, a Swiss unicorn pioneering digital therapeutics for neurorehabilitation and enhancement. MindMaze realized that by integrating an assessment with their intervention, as Dr. Kaplin did in the ECT study, they could achieve a better understanding of patient response and be able to respond appropriately.

Take Charge of Our Cognitive Health: DANA for Everyone

My vision for DANA has always been that every time you go to the doctor, in addition to taking your height, weight, blood pressure, and temperature, they will take your DANA brain vital. In 2016, I had the opportunity to bring this vision to the World Economic Forum in Davos, Switzerland. Although I had been to Davos before, this was my first time speaking on a main stage.

I started with a series of questions. "How many of you have ever misplaced your keys and couldn't find them as you were walking out the door?" About half the room raised their hands and there was some scattered laughter. "How many of you have ever forgotten someone's name whom you knew pretty well?" More hands went up and the laughter continued. "How do you know if this is normal, or if you have early onset Alzheimer's?"

Silence.

"The answer is...you don't."

I went on to make the case that measuring a brain vital not only helps us to treat cognitive conditions but could potentially ameliorate and even avoid them. I also encouraged them to remember key lifestyle interventions we can make to keep our brains healthy. The brain needs oxygen (exercise), nutrition (good food), and novelty (learning something new!).

After the session, I packed up my stuff and started to leave when a gentleman caught up with me. "I loved your presentation—you spoke with such passion! I've always been fascinated with neuroscience." I glanced at his nametag and realized why this stranger looked so familiar. It was Yo-Yo Ma!

He invited me to speak at an upcoming arts conference he was organizing at the Kennedy Center, and I walked out of Davos more motivated than ever to convince people to rethink brain health.

BRAIN POWER

We will all experience times in our lives when our cognitive function is negatively affected. It might be short term, due to medication or a treatable condition like sleep apnea; or more sustained, maybe due to an injury such as concussion or stroke; or a chronic condition resulting from stress or a disease state like multiple sclerosis; or part of a terminal condition such as dementia. And all of us, if we live long enough, will see our cognition decline naturally. That's why I believe brain vitals are so important to everyone's well-being.

When measuring your brain health becomes second nature—as common as checking your blood pressure—it will empower everyone, no matter their age, to spot changes sooner and take action.

In the case of my father, we still can't be sure if his cognitive decline was due to diabetes, not using his CPAP machine, cardiovascular disease, or the beginning stages of Alzheimer's. He ended up in long-term care when my mother could no longer manage his physical needs, but she called him every day and held weekly Zoom calls for the whole family.

When my mother died of COVID-19, we weren't even sure if he understood that she was gone. Yet he died suddenly, just thirty days later. I believe the cause was a broken heart.

The Big Island. (Photo Courtesy of AnthroTronix.)

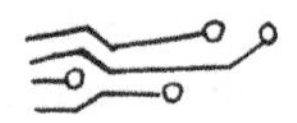

INSPIRATION AND REFLECTION #6
From Telemedicine to Telehealth

One of AnthroTronix's first projects, back in 2000, was to support the Strong Angel international disaster response preparedness project. As part of a telemedicine demonstration, we used portable satellite dishes (fairly new tech at the time) to help establish a medical communication infrastructure at a mock refugee camp on the Big Island of Hawaii. The goal was to demonstrate the potential of using telemedicine in a situation where doctors might not be able to treat refugees (or other vulnerable populations) in person.

One of the cornerstones to this effort was the Telemedicine Instrumentation Pack (TIP), NASA's first space-certified telemedicine system that had flown on the Space Shuttle *Endeavor* mission in 1989. TIP was basically a suitcase that included the electronics and telecommunications infrastructure to transmit data from the medical equipment also inside. This equipment included specialized medical scopes for visual inspection, an electronic stethoscope, a pulse oximeter and heart monitor, and a video camera.

I was one of the first guinea pigs for the demonstration and played the role of a refugee going through withdrawal symptoms from a heroin addiction. In this scenario, the doctors would need to distinguish the trauma from the refugee crisis and potential PTSD from the additional impact of drug misuse and withdrawal. I channeled my college acting experiences and prepared myself to act the part. To indicate track marks, we even drew lines on my arm with a pen. I was ready.

Our mock doctors were actually a team of real doctors in Chicago who were participating in Strong Angel. We initiated a video conference, and they looked very official and intimidating in their white coats. It wasn't hard to feel and act vulnerable in my grubby, dusty clothes, surrounded by technicians hovering with scopes and cameras from TIP. I think I even managed to cry. I may have been too convincing because the remote doctors became extremely concerned and wanted to help resolve my symptoms. They were clearly confused and at one point they even said, "Why are there pen marks all over your arm?" At that point, I had to drop the facade and remind them this was all fake—not just the pen marks! We moved on to a very earnest discussion about the types of disaster scenarios that these types of exercises could prepare doctors for.

Fast forward twenty years to 2020. Some clinicians had started teleconsulting with each other in specialty areas such as teleradiology or teledermatology, but from the patient perspective, telemedicine had barely changed. Very few consumers had experienced a tele-visit with their provider through video or phone call. Barriers included cross-state licensing limitations, reimbursement disqualifications from medical insurance, and just old-fashioned views about healthcare. Once you get sick, you go to the doctor, the doctor gives you orders, and you go home and sometimes comply. Diagnosis and treatment.

What about wellness and prevention? Tele*health* instead of tele*medicine*? This is really hard if we don't see our doctor regularly, and most of us don't. This is really hard if health insurance won't pay for tele-visits, and most of them didn't. This is really hard if we don't have the ability to self-monitor our own health, and—wait, we can! Over twenty years, the ability to monitor and treat our own health in the home had exploded. We now had 5G internet, smartphones, and home treatment systems for sleep disorders, diabetes, and other conditions. The capability for telehealth was there, but the healthcare system was too entrenched to change.

Then the COVID-19 public health crisis hit and we made twenty years of progress in less than two years. For example, before the pandemic, less than 1 percent of Medicare visits were telehealth based. During the pandemic, nearly half of them were.[35] Surveys suggest that both providers and patients want this trend to continue. Not surprising, as the need for telehealth has also driven growth. Baby boomers, the largest generation of people alive, are demanding new methods of communication with their health professionals so that they can stay healthy and age in place. To bring this full circle, mobile device-based software applications, like the DANA Brain Vital, for home care self-assessment and remote monitoring, may now find their place in this new world of telehealth.

---

35 "HHS Issues New Report Highlighting Dramatic Trends in Medicare Beneficiary Telehealth Utilization amid COVID-19," https://www.hhs.gov/about/news/2020/07/28/hhs-issues-new-report-highlighting-dramatic-trends-in-medicare-beneficiary-telehealth-utilization-amid-covid-19.html.

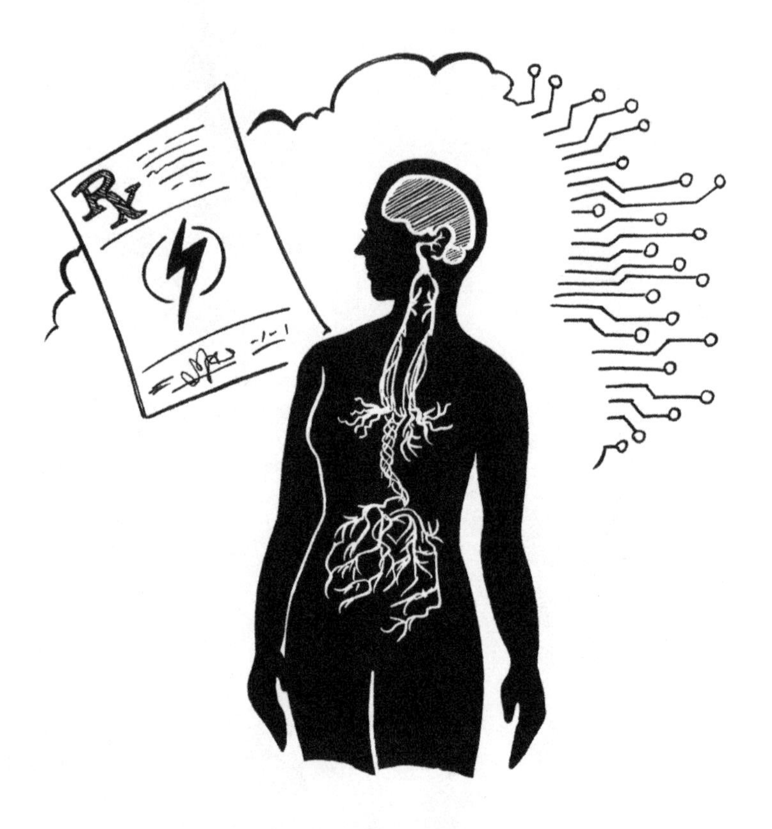

IF you find others who share your vision,
THEN you can realize a bigger impact.

CHAPTER 7

Electricity as Medicine

THE STORY OF STRING

IT'S JUNE 2019. My colleagues and I sit on one side of a long table in the sixteenth-floor conference room of Northwell Health, the largest healthcare provider in New York. We've partnered with Northwell to pitch a new product that uses electricity instead of drugs to treat disease.

Our investors are convinced of the potential for String, as we're calling our device, to reduce inflammation for people noninvasively, and they have already funded a prototype. Now we are talking to them about an ambitious crowd-funding effort. Project HeartString is our plan to manufacture thousands of String devices, distribute them for free, and crowdsource the data to find the best electrical signal to treat various inflammatory issues such as migraines, or chronic pain, or depression.

The investors are a bit dubious. Distributing so many devices for free is expensive, and crowdsourcing involves a lot of unknowns. The data can be messy. Trying to think through the questions, my gaze drifts to the clear blue sky outside. Suddenly, two heart-shaped balloons on a string float past the window. “Look, everyone!” It’s Pride Month, and the balloons have floated up from a celebration down below.

You can’t make this stuff up! We all gaze at this symbol of hope, connection, and amazing timing. Our investors take a leap of faith in crowdsourcing, and Project HeartString gets the green light. But why were we so confident about the product itself? Let me back up...

Serendipity

Sometimes your biggest opportunities come from unexpected sources.

Martine Rothblatt is one of the most amazing people I know. For a start, she founded Sirius XM radio. XM is now pervasive, but when Martine had the idea in the 1980s, no one could conceive of why you would want to listen to the same radio station wherever you were. Long-haul truckers might appreciate it, but general consumers? Now it’s standard in almost every car.

Martine then went on to found United Therapeutics, a pharmaceutical company focused on pulmonary arterial hypertension (PAH), to save her daughter’s life. At the time, a diagnosis of PAH was a death sentence, but Martine was able to track down an orphan drug, license it, test it, and get it through the FDA. In addition to making this lifesaving drug available to her daughter, Martine’s company helps thousands of others living with PAH.

Martine also happens to be transgender and transitioned while running Siriux XM. She is an incredible advocate for women and LGBTQ+ inclusion.

So in my mind, Martine is larger than life.

In 2003, AnthroTronix moved to downtown Silver Spring, where a few years later, Martine broke ground for a new headquarters. Eventually, this stunning, environmentally innovative United Therapeutics campus was completed right across the street from our building, so in a sense, we became neighbors. Over the next few years, many folks asked if I knew her and I always responded, "Of course I know *of* her, but we've never crossed paths."

One day, a mutual colleague said, "You two should know each other," and shot off a quick email saying just that. No surprise, she didn't respond at first. We are all busy, and a friend's opinion isn't necessarily a call to action.

I really didn't have a business reason to connect because in my mind, I worked in biomedical software and she worked in pharmaceuticals. Different worlds. But I did admire Martine and wanted to acknowledge the email. So I sent back a quick response. "Hi, Martine—I've long admired your work and wanted to particularly thank you for your advocacy, as I have a daughter who is part of the LGBTQ+ community."

I received a response in thirty seconds. "Would love to meet you, call my office and set up a time for us to meet."

Because of our schedules, the first slot that worked for both of us was three months out. By then, a coffee date was the last thing on my mind. My mother was ill with stage 4 lung cancer and I had been living in sweatpants, basically at the hospital with her all week. But I put on my game face and went to meet Martine.

It was one of the most interesting and impactful hours of my year!

Martine told me about some projects that were not necessarily on her website or part of UT's core mission, including one that grabbed my attention. She had become fascinated with Dr. Kevin Tracey's work on inflammation and the vagus nerve, which had been highlighted in a 2010 *Scientific American* article.[36] Dr. Tracey, President of the Feinstein Institutes for Medical Research at Northwell Health, had invented an implantable vagus nerve stimulator. Much like how a pacemaker electrically stimulates the heart, it stimulates the vagus nerve to reduce inflammation. Martine asked if AnthroTronix wanted to work with her and Dr. Tracey on a noninvasive version of the device. Cool project with cool peeps. Yes!

The Vagus Nerve and Inflammation

Dr. Tracey's lab at the Feinstein Institutes spent two decades proving that the inflammation reflex was controlled through the vagus nerve. This was revolutionary at the time as medicine had always treated the nervous system as separate and distinct from the immune system. But Dr. Tracey's team showed that electrical signals travel through this nerve from the brainstem to the spleen and then convert to a chemical signal, the neurotransmitter norepinephrine. Norepinephrine tells nearby immune-system T cells to release a second chemical messenger, acetylcholine. Acetylcholine binds to receptors, shutting down the tumor necrosis factor and other factors that produce inflammation.

Why does this matter? Inflammation underlies virtually all diseases, including cancer, diabetes, depression, obesity, and migraines, to name only a few. Right now, we treat inflammation with chemicals to mimic the effect of the neurotransmitter chemical response. For example, I might take an ibuprofen or other anti-inflammatory medicine for my swollen knee.

The problem is that those chemicals act on our whole body and are rarely targeted to where the inflammation is occurring. This can

36 Kevin Tracey, "Shock Medicine," *Scientific American* 312, no. 3 (2015), 28–35.

result in side effects like thinning of the blood (which is why you can't take ibuprofen before surgery) or nausea. Or in the case of other chemical treatments like chemotherapy drugs, much more serious side effects such as anemia, fatigue, infections, and loss of appetite can occur.

Dr. Tracey's work highlighted the importance of the nervous system in regulating the inflammatory response, or inflammatory reflex, just as it controls heart rate and other vital functions.[37] The inflammatory reflex maintains homeostasis in the immune system and maintains a healthy set point.

Furthermore, Dr. Tracey saw the potential for using electrical stimulation to mimic the electrical signals our body sends out, and to modulate them to respond in a healing way—reducing inflammation. This is called neuromodulation, and it can change the way chemicals are released in our body, freeing us from the harmful side effects of ingested medications and potentially reversing the disability that accompanies many inflammatory diseases.

Initially, Dr. Tracey envisioned an implantable device. He founded a company called SetPoint and began working toward FDA approval. But he grew frustrated with the time it was taking and how few people he was able to reach. SetPoint's clinical trials were approved to test only tens of patients when he could be helping hundreds or thousands.

This was brought home to him in a big way when he received an email from Kelly Owens, a twenty-eight-year-old who had just gone through one of the SetPoint trials in the Netherlands for patients with Crohn's disease. Since the age of fifteen, Kelly's Crohn's and inflammatory arthritis had left her with periods of immobility and she used a cane she named Rosie.

---

37 Kevin Tracey, "The Inflammatory Reflex," *Nature* 420, no. 6917 (2002): 853–59, https://doi.org/10.1038/nature01321.

Kelly was a go-getter. Born and raised in New Jersey, she completed her college degree in English literature and secondary education, married her sweetheart, and was living her best life. But she knew things could be better. When she heard about the trial in the Netherlands, she and her husband sold everything and moved to Amsterdam. She enrolled in the study and in 2017 had the SetPoint stimulator implanted in her neck. Every day she activated the electrical stimulation. She named her device Murph. Within three months of daily "murphing," her inflammation subsided remarkably. She could move almost as if she'd never had the disease at all!

She sent an email to Kevin: "You saved my life." As Kevin tells it, this email was buried in between a budget email from the institute and a reminder for a lab meeting. When he clicked on it, her words changed his life too. He decided to switch gears and focus on a way to get this life-changing technology to everyone.

But first, he hired Kelly! She is now the patient coordinator at the Feinstein Institutes, interacting with hundreds of patients looking for treatment for themselves or family members and facilitating patient recruitment for lifesaving clinical studies. And Rosie the cane sits unused in Dr. Tracey's office. A gift from Kelly.

A Meeting of the Minds

Martine said, "Kevin, you need to meet Cori," and "Cori, you need to meet Kevin," and when Martine speaks, we listen! Coincidentally, Kevin and I had recently spoken at the same conference but hadn't met. But that provided a nice point of connection and, in my mind, some validation, as I was a bit intimidated to meet the "father of bioelectronic medicine." Unfortunately, over two decades of experience had made me wary of older males in academic medicine as they did not always treat me as an equal. As it turned out, I shouldn't have been nervous. From our first phone call, Kevin and I were simpatico.

Kevin told me about his lab's push toward a noninvasive approach to neuromodulation. I told him about our journey with the DANA Brain Vital, and he immediately understood the importance of tracking brain health across the life span. He also realized DANA would be a great outcomes measure for vagus nerve stimulation, because reduced inflammation and pain should result in better attention and processing speed.

We agreed to join forces. United Therapeutics gave us funding and ninety days to build a prototype for a noninvasive transauricular vagus nerve stimulator, or taVNS, device. That's a mouthful, so let's break it down.

NONINVASIVE

In other words, you don't need surgery. You just apply the stimulation from the outside of the body. Many of us have experienced electrical stimulation, maybe to alleviate muscle pain at a chiropractor's office or to help strengthen our muscles at a physical therapist. Some may also have used a transcutaneous electrical neural stimulation device, commonly called a TENS unit, over the counter for muscle aches or relaxation.

TRANSAURICULAR

The meaning of "trans" is "across." And "auricular" refers to the ear. So just as transcutaneous stimulation with a TENS unit means across the skin, transauricular means across or through the ear.

Why the ear? Well, it turns out that through a fluke of evolution, the cymba concha part of the ear is an access point to the vagus nerve. This was discovered by researchers at Feinstein and others. You might ask, "Where is the cymba concha?" Put your pointer finger in your ear as if you were trying to block out sound. Leave it there and put your middle finger in the space above it. *That* is your cymba concha.

VAGUS NERVE STIMULATION

Essentially, what Kevin and Martine were suggesting was a TENS device for your vagus nerve. However, the device we wanted to create was different from a TENS unit in some key ways. One was that the physical profile for a vagus nerve device would likely be different than a TENS unit, which has sticky pads and wires. Another big difference was that to optimize the treatment, we wanted a responsive device, one that could measure the body's physiology or response to stimulation, and change the level of stimulation accordingly.

Building a Prototype

Martine had funded a "bench" prototype, also known as a "looks like feels like" prototype—it demonstrates the look and functionality of the final product but is not actually the final version. Our vision was for something easy to use that could be part of an everyday routine. We didn't want it to look like a medical device but more like a cool accessory so people would want to wear it while jogging, listening to music, meditating, or just thinking.

So for the bench prototype, we bought some Beats headsets. We embedded carbon fiber electrodes into the earpieces, which would sit in the cymba concha when the earpieces were in place. We covered the electrodes with sponges so that when the user applied a little saline or salt water, the electrodes would make good contact with the skin.

We developed an app that not only used our DANA Brain Vital but also asked about your mood, stress, sleep quality, and levels and locations of pain. After collecting this information, you could control the String prototype through the app. Our prototype delivered a five-minute "dose" of electrical stimulation to the vagus nerve—and you could listen to great music too! The idea was that as more and more data was collected over time from many individuals, the app would learn to optimize the dose for each individual and across diseases.

A Short-Lived Company

Ninety days after my meeting with Martine, Kevin and I had a prototype. We met on the sixteenth floor of Northwell to pitch myString to Martine and her team and to witness the double heart balloon floating past the window. Martine agreed to invest further in Project HeartString, and we started the company in the fall of 2019.

By January 2020, myString had a CEO, interest from major investors, and a tentative deal with a group based in China, among other locations. Once the executives were back in the office, after the Chinese New Year in late January, we could hammer out the details.

You can probably guess what happened in February.

COVID-19 hit, flights were canceled, and our potential investors had more pressing concerns, like making sure their families were safe and protecting existing investments. A few months later, with the world in quarantine, we evaluated our options and decided to close down the company.

It's Not the End but a New Beginning

We were not the only ones with a vision to bring electroceuticals to patients in need. Pharmaceuticals are a trillion-dollar industry; electroceuticals could be big business too. Several companies were seeking FDA clearance for electrical stimulation devices designed to treat epilepsy, depression, obesity, addiction, and more.

One company in particular was intriguing. Cala Health had received FDA clearance for an electrical stimulation device, the Trio, worn on the wrist to help reduce or eliminate essential tremor, including for Parkinson's patients.

Kevin knew that Cala was interested in vagus nerve stimulation as they had licensed technology from scientists at Harvard.

Now that myString was no more, he connected me with Cala's founder, Kate Rosenbluth. Kate convinced me that Cala's approach of making a vagus nerve stimulator a prescription device was more affordable than our direct-to-consumer plan, with the potential to reach many patients who needed it. She asked Kevin and me to help with their product path to market.

Fast forward a few months. We're back at Northwell Health, this time on the first floor of the Feinstein Institutes building, looking out the window on a very stormy day.

I'm sitting with Kevin, Kelly Owens, and several others from the Northwell team. Kate and some folks from Cala Health have joined by Zoom and are on-screen in front of us. The goal of the meeting is to plan the study that will launch transauricular vagus nerve stimulation. I can't divulge the details, but stay tuned, because balloons or not, we are going to help thousands of people!

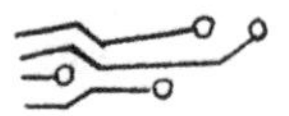

INSPIRATION AND REFLECTION #7
Electroceuticals, or the Healing Powers of the Body Electric

Our body runs on electricity. But electricity gets a bad rap. We think of accidents involving electricity, like being electrocuted or struck by lightning, or imagine frightening scenes of electroconvulsive therapy from *One Flew over the Cuckoo's Nest.*

However, we can tap into that electric current for good. This is the basis of bioelectronic medicine, the idea of treating disease with electrical signals as an alternative to chemical signals or drugs. Electroceuticals instead of pharmaceuticals. Electricity has actually been used in this way throughout history. Roman court physician

Scribonius Largus had patients stand on a bioelectric fish to cure headaches and treat gout. More recently, electroconvulsive therapy saved the lives of people with severe depression, with far fewer side effects than many imagine.[38]

Very simply, electroceuticals are devices that treat ailments with controlled electrical impulses by targeting the nervous system. Today, the electroceutical market is growing fast, with many types of stimulators for a wide variety of conditions. The first Transcutaneous Electrical Nerve Stimulation (TENS) unit was patented in 1974 by Medtronic. Initially, it was used to test a chronic pain patient's pain tolerance before getting an electrode implant. To everyone's surprise, after using the TENS unit, the patients had enough pain relief that they no longer needed the implant! Thus was born the modern electroceutical. The use of TENS units has continued, and today thousands of people buy personal units for their pain management yearly.

Although they are the most widely known, TENS units are only one part of the electroceutical market. At the other end of the electroceutical device spectrum are implantable devices.

Implantable electrical devices, like a pacemaker for regular heart function, have been around for over fifty years. In recent decades, implants for deep-brain electrical stimulation for Parkinson's patients have emerged. These send electrical impulses to the area of the brain that controls motor function in order to stop tremors. More recently, the FDA has approved implantable electrical stimu-

---

38 K.E.D. Coan, "Electroconvulsive Therapy Is Safe for Treatment of Mental Conditions, Shows Large-Scale Study," *frontiers Science News*, November 24, 2021, https://blog.frontiersin.org/2021/11/24/electroconvulsive-therapy-is-safe-for-treatment-of-mental-conditions-shows-large-scale-study/.

lation devices for bladder and bowel control and for direct spinal cord stimulation to relieve pain.[39] And of course, in Chapter 8 I talk about the SetPoint device, which was developed to treat rheumatoid arthritis, Crohn's disease, and other autoimmune disorders.

Electroceuticals are turning up everywhere in medicine for three reasons. First of all, backlash against the pharmaceutical industry is increasing due to concerns about side effects, high costs, and problems like the opioid epidemic. Second, in the wake of the COVID-19 pandemic there has been an increasing demand for at-home medical solutions, including remote monitoring and treatment. Finally and most importantly, the science underpinning electroceuticals is growing. Our understanding of the nervous system and our ability to use neuromodulation to treat disease has incredible potential.

These technologies are only the beginning. Experts predict the global electroceutical market to grow from US$16.8 billion in 2021 to US$21.5 billion by 2026! Widespread use in the United States will be driven by the availability of noninvasive, scientifically based FDA-cleared devices that are coming to the market as we speak. As the research continues and our ability to refine and target known neural circuits improves, the opportunities for treatment will be limitless.

39 "InterStimTM II System," MedTronic, last updated February 2022, https://www.medtronic.com/us-en/healthcare-professionals/products/urology/sacral-neuromodulation-systems/interstim-ii.html.

IF we embrace technology,
THEN we become more than the sum of our parts.

CHAPTER 8

Enhancements for Humans

MAKING THE CASE FOR CONNECTED COGNITION

Welcome to the Kennedy Center Arts Summit session on Art and Science at the Edge: Innovations That Create New Culture. My name is Dr. Cori Lathan and I'm a technology entrepreneur and neuroscientist. I also co-chair the World Economic Forum's Council on Human Enhancement and Longevity.

It's 2017 and I'm addressing an audience of scientists, artists, and business leaders about a topic near my heart.

So what is human enhancement? We tend to think of it at the individual level. For example, I'd like to replace this knee with a better one, fix my aging eyes, and maybe use biofeedback to improve my mental and emotional state. But our council's definition of human enhancement is more than just empowering an individual—it also has to make communities more inclusive, cohesive, and resilient. In other words, human enhancement is a collective experience.

One of the themes of today's conference is achieving cultural moonshots. Cultural moonshots are audacious and risky innovations that revolutionize the way we collectively view art, science, and each other. My cultural moonshot is that by 2030, we will be regularly enhancing our understanding of each other and communication with each other, through real-time data from wearables and sensor technology.

Does this kind of communication sound crazy? Every project I've told you about—plus a couple I'll get to in this chapter—have taught me it can be done. But to get there, each of us needs to embrace human enhancement technology—including you! What do I mean by that? Let me back up...

What Do We Mean by Human Performance?

Think of yourself as an information processor. You take information in from your senses, process it, and then act on it. For instance, if you see that the hallway is dark, you might turn on the light. Your actions change the information coming to your senses—now you see that the hallway is light and you move forward until you see something that requires a new action—and the loop continues. In engineering, this is called a closed-loop system.

An open-loop system is one where you keep acting without seeing the consequences. Heating soup on the stove can be open-loop—especially if you walk away like I do! You turn the burner on and it stays on even if the soup boils away. You are not there to observe and make changes to the burner.

In terms of my work, a soldier controlling a robot to go into a building is a closed-loop application as you base your control inputs on feedback you get from the robot. Repeatedly presenting stimuli to astronauts through their headsets and then having them react is an open loop. The process is great for collecting data, but it doesn't, in that moment, change or improve the astronaut's performance.

We can enhance human performance anywhere along the continuum of a closed or open loop—by improving access to information, improving our processing, or improving our actions.

ACCESS TO INFORMATION

Think of all the ways we use technology to improve our access to information. Wearing glasses enables clearer vision. Wearing a hearing aid improves hearing. Simple tools like binoculars or remote cameras extend visual access beyond what we'd consider normal human capability. As technology advances, it can provide access in ways that we hadn't previously conceived of. For example, the smartphone has put more information at everyone's fingertips than a president of the United States could access just fifty years ago.

One of the innovative ways AnthroTronix improved access to information for soldiers was through vibrotactile, or haptic, feedback. When you feel your phone vibrating to alert you to a message, that's a vibrotactile alert. Our technology enabled a soldier to control a robot purely through this type of feedback by equipping the robot with an ultrasonic sensor array. During another military-funded program, we used this haptic feedback for room-clearing training. To clear a room, a soldier needed to sweep the whole space; the vibrotactile array alerted them if they missed any areas.

We even used haptic feedback for communication between soldiers. We built a vibrotactile array into a vest that soldiers could wear and developed patterns of code they could use to communicate while patrolling. Presenting the soldier with haptic instead of visual cues enabled them to communicate faster, which is critical in any crisis scenario.

PROCESSING

After access, we have the *processing* part of the loop. Once information is received, in whatever form, it must be interpreted. Our brain processes information from multiple senses, accesses short-term and long-term memory to figure out what to do with the information, and then decides to act on that information in any number of ways.

As a neuroscientist, I'm deeply invested in understanding and improving the brain's cognitive abilities. But before we dig into that, let's talk about action.

ACTION

Action can be a one and done, like composing an email, shouting out loud, or reaching for an object. Or it can be part of a larger loop—one step in controlling a robot as it moves down a hallway, or taking a bite of food during lunch, or putting the next foot forward while running.

We have thousands of tools that help us complete or enhance the action—a keyboard or microphone, a megaphone, or a pair of tongs. You might use a low-tech tool like a fork or chopsticks to eat, or maybe a high-tech prosthetic lower limb for running. In the case of controlling a robot, that tool might be a joystick or an instrumented glove.

All these examples enhance human performance or capability without changing what it means to be human. But what if we start using technology to extend so-called normal human capabilities beyond what is typical?

It's Time to Talk about Processing

Enhancing human capability was the goal of a program we worked on with DARPA in the early 2000s. The Augmented Cognition program (AugCog) began in 2001 and ended with the launch of the

Augmented Cognition International Society, which held its first conference in July 2005.

The field of augmented cognition is based on the premise that you can objectively measure, primarily through physiological measures, the cognitive state of a system operator. Basically, you can measure how hard their brain is working; then you can change the system based on the person's functional state to get more work done. In other words, reorganize the inputs so that the brain can process more stuff.

To test this, DARPA's AugCog program funded experiments in a range of computer-based tasks. Many were military simulations, including flight, driving, and ship-based operations, as well as dismounted warfighter tasks like robot operation. The program had the ambitious goal of measuring someone's workflow on a computer and improving it by as much as 500 percent.

MEASURING COGNITION

One of the many metrics AnthroTronix explored for AugCog was working memory. Working memory is generally defined as the number of items you can hold in your memory at once. The average is about seven. If I tell you seven items, you can repeat them back with a high degree of accuracy, but if I give you eight or more, you are more likely to make mistakes.

Of course, this number depends on lots of things, including how the information is presented and what else you are doing that might distract you. So we designed experiments that tested the working memory of a naval missile monitor—in other words, a specialist using a computer system to monitor the status and control the activity of naval missiles.

First, we established a baseline for the operator's cognitive state while they were at the computer. We used electroencephalography (EEG) to measure brain activity, electrodermal activity (or skin conductance) to monitor stress levels, and electrocardiography

(ECG) to record heart activity. We also established the number of missile targeting tasks the participants could manage at one time given the current system.

We then tried to predict an increase in working memory workload and present the information differently. The EEG in particular was useful in determining which parts of the brain were active, indicating how we might present the tasks in a different way. For example, if too many tasks were being presented visually, we presented the task auditorily, or we might change the order or pacing of the presentation of tasks. Using this method of "intelligent sequencing," participants increased the number of tasks processed by 642 percent!

OPERATIONALIZING IT

Our work at AugCog was at the forefront of "operational neuroscience," a term coined by another DARPA program manager, Dr. Amy Kruse. Augmented cognition uses neuroscience to improve existing human–computer interfaces and therefore human performance—thus operationalizing the science. In fact, Amy argues that applying neurophysiological measures in functional environments is a field in and of itself.

Amy's program at DARPA, Neurotechnology for Intelligence Analysts, took the idea of operational neuroscience in another direction. We know that the brain processes a lot more information than we are aware of. What if we could tap into those subconscious signals to help us analyze information?

For example, when shown a series of images and told to find the truck, you have to perform a *Where's Waldo*–like visual search. An EEG measures each brain response, called an event-related potential (ERP), even before you recognize and point to the truck. ERPs are why we're able to test things like a newborn's hearing. We can record their brain signal while audio clicks are presented to each ear. The infant obviously can't tell you, "Yep, I heard that click in my left ear," but their brain signals the processing of the information.

So in the example of the analysts, very large images that would have taken analysts many minutes to scroll through manually were segmented into small images and were presented in a matter of seconds. Any image that evoked an ERP could be put aside for further examination. Imagine the next generation of TSA agents connected to EEG caps or other monitoring devices that automatically alert them if their brain recognizes an object of interest. Those airport security lines would move a lot faster!

THE POSSIBILITIES ARE ENDLESS

Once I started thinking about neuroscience as an operational proposition, I realized that there could be many ways to capture the power of our brain's ability to process information to enhance human performance.

For instance, what if we could operationalize the "cocktail effect"? This effect is the phenomenon where you are in a group of people with many conversations going on that you aren't paying attention to, but if you hear your name, you immediately register it and are on alert. Using operational neuroscience, what if you could train your brain to alert you to not only the sound of your name but other names you care about, like friends or family? Or important words like "chocolate"? Or the sound of a particular person's voice?

I was recently at a party and my friends couldn't get my attention—apparently, I was too engrossed in conversation for even the "cocktail effect" to work. One of them had an idea. "ROBOT!" I immediately turned and looked. So apparently, my brain has already been rewired!

In 2010, I learned about the work of Dr. Uri Hasson, a neuroscientist at Princeton University. Uri used EEG and fMRI data to show that when someone is talking to you and you truly understand and comprehend what they are saying, the two of you are literally on the same wavelength. Neurologically, your brain waves synchronize.

I found this absolutely amazing. It means that when you sense you have a connection with someone, you actually do—and you can measure it! And if you can measure it, you can probably capture it and manipulate it. I found myself more and more drawn to ideas and projects that would enable me to further this line of work.

The Case for Connected Cognition

Around the time I learned of Dr. Hasson's work, I was asked to join DARPA's Information Science and Technology (ISAT) committee. ISAT is a group of scientists and engineers tasked with doing deep dives into topics that might be of interest to DARPA. Rarely did an ISAT topic become a full DARPA program itself, but almost every topic sparked ideas that were incorporated into other programs.

The ISAT committee was made up overwhelmingly of computer scientists, but that year they appointed two neuroscientists: myself and Dr. Todd Coleman. Todd directed the Neural Interaction Laboratory at the University of California San Diego for many years and then moved to Stanford University.

At our first meeting, Todd and I discovered we were both interested in wearable tech. At the time, he was working on wireless epidermal electronics that conformed to the body of the user and picked up EEG and EMG, electromyography or muscle activity. I was fresh off the AugCog program and thinking about operational applications of neural signaling and data collection. We began discussing the concept of enhancing human-to-human connection using wearable technology.

Todd and I proposed the study topic "Cognitive Coupling" to explore the synchronization of neural signals between people. We were interested in the idea that if you capture these signals and connect them to each other through computer technology, you have *connected the crowd to the cloud through neural signals.*

In essence, we wanted to operationalize the neuroscience of neural synchrony. Our study identified three areas that needed more attention, all centered on enhancing the brain–computer interface:

1. Neural coupling for expert amplification. Can we understand what it means to be an "expert" and give biofeedback to a "novice" so that they more quickly become an expert?
2. Neural coupling to accelerate team training and performance. Can we develop metrics that demonstrate a team's readiness to perform and get them to that point more quickly?
3. The neuroscience of interaction. We need to invest in low-cost portable ways to measure the brain, and we need to understand how to influence neural signals.

Over the next five years, DARPA launched multiple programs in the area of brain–computer interfaces. One of our study participants, DARPA program manager Bill Casebeer, went on to run several of these. His work is a good example of the influence our study had.

Dr. Casebeer was a lieutenant colonel in the air force, a former philosophy professor at the US Air Force Academy, and a neuroethicist. With that background, it may come as no surprise that some of the DARPA studies he ran were quite novel, like Narrative Networks, which looked at how our brains react to stories and then tried to create new stories that could help reduce violence and build trust.[40] He also spearheaded the low-cost EEG effort that brought

---

40 Sharon Weinberger, "The Pentagon wants to understand the science behind what makes people violent. The question is what do they plan to do with it?" *BBC*, November 18, 2014, https://www.bbc.com/future/article/20120501-building-the-like-me-weapon; "Narrative Networks," Our Research, DARPA, accessed March 18, 2022, https://www.darpa.mil/program/narrative-networks.

brain–computer interfaces to the maker community.[41] The idea was that if we could provide access to affordable EEG recording systems to the broader community of both professional and amateur neuroscientists, we could advance the brain–computer interface field in innovative ways.

Another one of our study participants was Dr. Ray Perez, a program officer at the Office of Naval Research, a top government funding agency. Dr. Perez funded one of my colleagues at AnthroTronix, Dr. Anna Skinner, and she became the principal investigator for the pilot project: Identifying Neural Synchronies of Peer Interaction in a Real-Time Environment for STEM (INSPIRE STEM).

Our ISAT study had proposed that neural coupling could enhance skill acquisition. The hypothesis was that the more an instructor and their student are neurally coupled, or "on the same wavelength," the quicker the student would reach proficiency.

Dr. Skinner put this hypothesis to the test in a controlled environment of the spatial reasoning video game Tetris. Tetris was selected based on its requirement of speed, decision making, and problem-solving skills. The goal of the pilot study was to provide groundwork for discovering neural synchrony measures to drive more effective computer-based instruction for STEM.

Tetris "experts" provided verbal coaching to Tetris novices during the game play while EEG measures of engagement and workload were recorded for both the expert and the novice. The initial results indicated that the EEG-based metrics of expert-novice neural synchrony were a reliable indicator of novice performance!

The potential impact of neural coupling is still a wide-open area of research, which many of our colleagues are still working

---

41 Kristen Butler, "DARPA pitches DIY brain-scanning at Maker Faire," *UPI*, September 23, 2013, https://www.upi.com/Science_News/2013/09/23/DARPA-pitches-DIY-brain-scanning-at-Maker-Faire/6101379952887/.

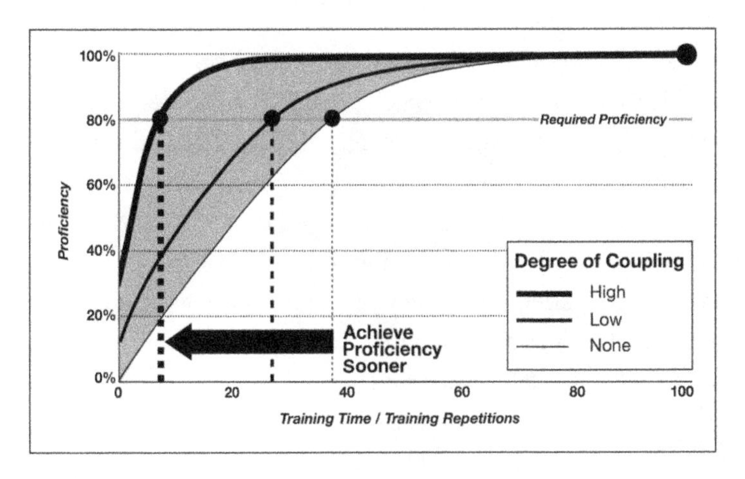

Proposed models of accelerated learning through using neural coupling.

on. But as our ISAT study was over, I had moved on to other questions, like whether people *want* to be enhanced and how far would they go.

Quantifying the Support for Human Enhancement

In 2016, I was asked to co-chair a new World Economic Forum council on human enhancement along with Dr. Linda Fried, the dean of Columbia School of Public Health.

The forum first welcomed me into the fold when they named AnthroTronix a technology pioneer in 2005 and then invited me into the 2006 Class of Young Global Leaders, which recognizes leaders under forty committed to driving "positive change in the world." Through the YGL community, I formed relationships with innovators in government, business, and the arts and saw firsthand the power of collaborating across disciplines and borders.

Co-chairing the council on human enhancement gave me the chance to advocate for human–technology interaction on a global scale. But I had been involved with other forum councils in the past and I knew it would be a challenge to get a group of high-powered individuals with very diverse backgrounds and agendas on the same page to identify a realistic deliverable, especially given that we were all volunteers with day jobs.

Our first meeting was in Dubai in the United Arab Emirates. I was immediately put at ease seeing some familiar faces. One of the committee members was Professor Daphne Bavelier, a French neuroscientist heading up a cognitive science department in Geneva. Daphne coincidentally also happened to be one of my best friends from graduate school! I was also excited to spend time with a fellow YGL, Mark Pollock, a resilience expert from Ireland and the first blind person to race to the South Pole. Another longtime colleague, Dr. Derek Yach, is a South African epidemiologist who convinced Philip Morris International to give over half a billion dollars to start a foundation to end smoking.[42]

As I'd anticipated, it took some discussion to agree on a task that could make the world rethink human–technology interaction *and* be completed in less than two years. Another council member, Debra Whitman, had an idea. Debra is the Chief Public Policy Officer at AARP, the largest nonprofit in the United States. As soon as you turn fifty, you can be sure that you will get your membership invitation in the mail. Their motto, which I love, is "Disrupt aging!"

Through Debra's leadership, we partnered with AARP for a project that would further their mission while enabling us to add to the body of scientific knowledge around human enhancement technologies. We laid out a plan to study public awareness, acceptance, and potential concerns around various enhancement technologies.

To gather data on what the public was feeling, we used a survey.

---

42 Disclosure: I'm on the board of The Foundation for a SmokeFree World (www.smokefreeworld.org).

OUR SURVEY SAYS

The AARP contracted with University of Chicago's AmeriSpeak Panel to survey a representative sample of 2,025 Americans aged eighteen and up. One of the survey's first objectives was to level-set and explain what we meant by human enhancement. Here is the definition we used: "Human enhancement refers to technologies that enhance certain aspects of human performance with the aim of improving quality of life at all ages and stages of life." Participants were then asked to consider existing technologies that already do this, such as pacemakers, medications, prosthetics, and joint replacements.

Next, we asked them to think about enhancement technologies a little differently. The survey presented a framework, which places enhancement technologies on a continuum. At one end is therapeutic use, for restoring what we'd consider normal ability. At the other end is technology that enhances ability far beyond what is considered normal. We wanted to know what they'd consider an appropriate level of human enhancement for various situations.

The questions covered several health issues, but I'll focus on three that parallel our model of the human as an information processor: vision enhancements to improve sensory information; cognitive enhancements for processing information, including medications and implantable devices; and joint replacements for acting on information.

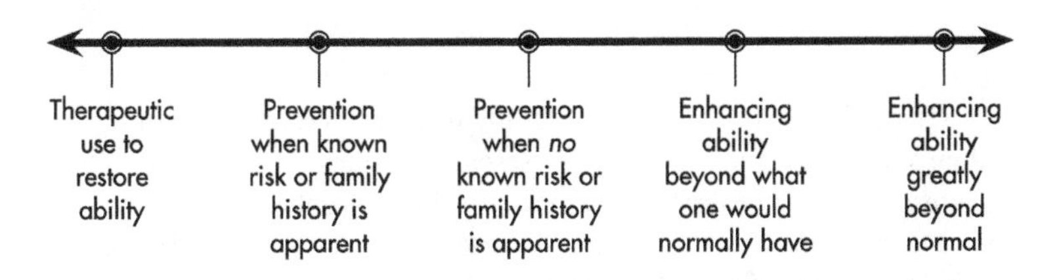

Technology enhancement continuum.

For the two physical health–related issues, vision and joint problems, there was almost universal agreement (over 95 percent) that enhancements are appropriate for restoring what we'd consider normal function or preventing loss of function. Interestingly, fewer respondents (88 percent) supported enhancements to improve therapeutic cognitive function.

In contrast, far fewer respondents supported enhancements where no need is evident or where they would go beyond what are considered normal human capabilities. Support for joint replacements dropped to 33 percent. Support for cognitive enhancements was close to that, with 35 percent supporting cognitive medication and 31 percent supporting cognitive implants. There was a bit more support for vision enhancements—44 percent thought those were fine even beyond normal therapeutic levels.

Interestingly, despite this generally low support for enhancements beyond normal, when asked if they were interested in human enhancement technologies to improve *their own cognitive abilities* beyond normal, 43 percent approved of a medication that could provide that—almost 10 percent more than had approved of it for others. The full results are available on AARP's website and a summary was published through *Scientific American* online.[43,44]

I am clearly part of the minority. I fully expect that as I age, I will have new opportunities to improve my brain function and prevent the loss of function—and I plan to take them. How about you? Would you have surgery to gain superhuman eyesight? How far

---

43 Debra Whitman, "U.S. Public Opinion and Interest on Human Enhancements Technology," AARP Research, January 2018, https://www.aarp.org/research/topics/health/info-2018/human-enhancement.html.

44 Debra Whitman et al., "What Americans Think of Human Enhancement Technologies," *Scientific American*, January 16, 2018, https://blogs.scientificamerican.com/observations/what-americans-think-of-human-enhancement-technologies/.

would you go to be smarter? Can I convince you to support those enhancements for all of us?

Innovation, Empathy, and Human–Machine Interaction

As I mentioned, Uri Hasson's work has shown that when you are truly engaged in a dialogue, whether as speaker or a listener, your brain and that of the other person are literally on the same wavelength—your brains synchronize. But it goes beyond that.[45]

Dr. Hasson shows that if you and the other person start a conversation after a shared experience as simple as listening to the same musical performance, your interpretations of events are more likely to be similar. The inverse is also true: if we start out in different places, it is less likely that we can find agreement. In fact, he extrapolates that given our current political climate, "we are in danger of losing our common ground with people who think slightly differently than us."

What if instead, we could use our neural connections to build empathy with and for each other? Think about our cocktail party example. What if everyone was recording their brain activity and uploading it to social media so you could tell who at that party was interested in the same things you were? Maybe it could be the basis of a new dating app...or a virtual chocolate lover's club...or an entirely new approach to discussing and resolving our political differences.

Where We Go from Here

I said that I hoped this chapter might inspire you to rethink your own relationship to technology and human performance. Yes, I'm

45 Uri Hasson, "This Is Your Brain on Communication," filmed February 2016, video, 14:42, https://www.ted.com/talks/uri_hasson_this_is_your_brain_on_communication.

deeply interested in technology that enhances individual human performance all along the information processing loop. But I believe that embracing the potential for neural interconnectedness could have a far greater impact on humanity.

Star Trek fans, I hear you. This level of neural connectedness may sound a little too close to the Borg Collective. The Borg were a human-machine hybrid who used neural interfaces to connect every member of their society to each other. But their approach was machine-centric, not human-centric. They obeyed a leader who ruled with complete authority and had no facility for empathy. They used the connection to control, not to listen.

But my vision for the future of human–machine interaction is human-centered. Humans plus technology should *always* equal more than humans alone or technology alone. This vision builds on innovation and empathy, natural human capacities that can be enhanced by technology—specifically, by capturing and communicating the neural connectivity we already share.

So instead of seeing human–machine interaction as a dehumanizing force, I challenge you to imagine a human-centered group consciousness. Imagine how your own opinions might shift if you actually shared other people's perceptions. What social problems might suddenly become crucial to solve? How might our empathy grow if we perceived the suffering of our fellow human beings on a neurological level?

Thinking a little further into the future, what problems might we avoid altogether by investing in connected cognition? There's a quote attributed to Einstein that inspires my enthusiasm for this unabashedly optimistic vision: "Intellectuals solve problems, geniuses prevent them." How would the next pandemic look if we all personally understood the impact of the illness on some individuals and the fears about vaccines on others? Or could we avoid another pandemic altogether?

If you think of collected consciousness *as an individual entity*, then it becomes in our own self-interest to fix the pain and suffering in the world, to optimize the potential of every human being. In a global consciousness, every problem is worth solving because every problem is shared.

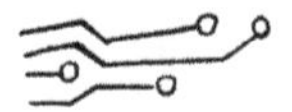

INSPIRATION AND REFLECTION #8
Harnessing Brain Power

In 2011, AnthroTronix partnered with two pioneering women in neurotech to try and take social networking to the next level using neurotechnology. Chris Berka, founder of Advanced Brain Monitoring, and Tan Le, founder of Emotiv, both invented devices that measure brain activity using EEG, or electroencephalography, which measures the electrical activity through the scalp. Together, we submitted a proposal to Breakout Labs, an organization started by Facebook's first investor to fund cutting-edge scientific research deemed too risky for the for-profit sector.

Our proposal was "Reinventing Facebook and the Future of Social Networking." We proposed to investigate neural and physiological markers of emotion by measuring EEG, heart rate, and skin conductance of Facebook users. We wanted to see if we could enhance users' experience of shared personal and emotional moments and create a more meaningful social networking space.

At the time, we were thinking about the future of brain–computer interfaces (BCIs) and how to use Facebook to engage communities on a deeper level. Little did we know that they would evolve to become Meta and that #neurotech #metaverse would become a thing.

Our proposal was rejected.

One reviewer said, "The focus on developing noninvasive measures to monitor coupling between brains during interactive behaviors could not be more timely," but *all* reviewers felt that the technology couldn't accomplish what we proposed.

Fast forward to the present day, and Emotiv's headsets have shown an incredible breadth of research applications for captured brain activity, including controlling a wheelchair, detecting work stress before it gets overwhelming, or using just your mind to draw pictures or create music. And beyond the reviewers' expectations, Emotiv and other consumer EEG products like Muse and DREEM have democratized brain–computer interfaces. For instance, DREEM uses EEG as well as movement sensors to track brain activity and sleep stages during sleep. It then uses that information to generate "pink noise" to keep the user calm and asleep. Muse is the leader in using EEG biofeedback for guided meditation, combining it with breath work and visualization. They have curated extensive libraries on their app to reduce stress, increase attention, improve sleep, and more.

A few years after our failed proposal to expand the Facebook interface, I wrote an article titled, "Will the Computers of the Future Read Our Mind?"[46] I suggested that in time, we will be able to use BCIs for simple tasks such as controlling the cursor on a computer screen or interacting, hands-free, with mobile phones. Or that we could soon have a computer that acts as an intelligent tutor, using a BCI to sense when your attention wanders and restructuring the lesson to set an efficient pace of study.

This was inspired by what was thought to be the first demonstration of a human-to-human brain interface, with one researcher at the

---

46 Corinna Lathan, "Will the computers of the future read our minds?" World Economic Forum, November 10, 2014 https://www.weforum.org/agenda/2014/11/will-the-computers-of-the-future-read-our-minds/.

University of Washington sending a brain signal, recorded through EEG, over the internet to control the hand motion of another researcher.[47] More recently, Facebook Reality Labs has partnered with academic researchers to develop a noninvasive silent speech interface for a future computing platform.[48] This is getting closer and closer to #neurotech #metaverse!

But in order to reach this vision of a brain-powered metaverse, we need more makers or "neuro-hackers" to get involved, especially ones who are focused on empowering the user and not optimizing Facebook ads. That's where OpenBCI comes in. OpenBCI was one of the low-cost EEG projects funded by DARPA to create open-source tools to create brain–computer interfaces. OpenBCI was also founded with the mission to develop computer interfaces in an ethical way, protecting user mental health and user agency, or feelings of control.

Now they have extended their EEG tools to include a broader platform. This new platform will allow us neuro-hackers to combine biometrics such as heart rate with brain activity and interface with mixed-reality environments. So as Facebook, or Meta, and others build out the metaverse, we will be ready to power it with the brain.

---

47 Doree Armstrong and Michelle Ma, "Researcher controls colleague's motions in 1st human brain-to-brain interface," *UW News*, August 27, 2013, https://www.washington.edu/news/2013/08/27/researcher-controls-colleagues-motions-in-1st-human-brain-to-brain-interface/.

48 Tech@Facebook, "Imagining a new interface: Hands-free communication without saying a word," *Inside Facebook Reality Labs* (blog), March 30, 2020, https://tech.fb.com/imagining-a-new-interface-hands-free-communication-without-saying-a-word/.

IF I inspire my daughters,
THEN they will change the world.

CHAPTER 9

Mind of a Maker

INSPIRING MORE STORIES

I DON THE VR headset and strap my fingers onto the digits of the robot arms. I place my feet on the rotating navigation disk and close my eyes. When I open them, I've been transported into a room with my colleague Jerry Pratt, who is sitting across the table. "Hi, Cori, great to see you! I'm so glad your robot avatar could join me!"

He reaches out to shake my hand, and as I extend mine, I can see my robot hand in great detail. I easily grasp his hand and reply, "It's amazing we can meet this way!"

"Would you like to join me in a toast?"

"Sure!" I reach for one of the two drinks in front of me, raise it to his, and we clink glasses. "Cheers!"

This is *not* science fiction. It's the ANA Avatar XPRIZE semifinal competition in Miami, September 2021. Jerry holds up a mirror and I can see a reconstructed image of my face smiling and laughing. I have two powerful and surprisingly beautiful robot arms. In this moment, I realize that I'm looking at the future.

Never mind that the task we've just completed—making a toast via avatars—is unlikely to be the killer app that drives the commercial viability of the technology. We aren't here to buy a product or solve a problem; we're here to be inspired. But how did I snag an invite to this virtual party? Let me back up...

Flexing the Maker Muscle

So far, every invention I've told you about has been a solution to a real-world problem. And that's an essential way to approach invention, particularly for eventual commercial success. But sometimes you have to invent or create just because you can! Inventing is not only an art—it's also a muscle.

Traditionally, not everyone has been encouraged to flex this "maker" muscle. As a young woman in STEM, I spent years being the only or one of the only women in classes, graduate research labs, and business meetings. But long before I started my company, I read something that dictated a path for my life and has equal importance as—and is, in fact, symbiotically intertwined with—my professional accomplishments.

The Problem: Implicit Bias

"If the cure for cancer was in the mind of a girl, we might never discover it."

—Myra Sadker

It was 1992 and I was a grad student at MIT, reading a report by Myra and David Sadker, titled, "How Schools Shortchange Girls." My female friends and I were long used to having to work harder for STEM opportunities—heck, I had to petition my high school just to take power tech with the boys instead of home economics with the girls—but Myra's quote hit me like a ton of bricks. The status quo was jeopardizing everyone's future.

One of the report's findings that hit hardest was from an elementary school study that recorded teachers' interactions with kids as they worked.[49] They identified four types of things teachers would say to kids, paraphrased here: (1) "That sucks," which rarely happened in the classroom; (2) "That's awesome!" which also rarely happened in the classroom; (3) "Fine," "Good," "Yes," which were the majority of interactions with girls; and (4) "Good. And why don't you try this?" or "What do you think about that?" which were the majority of interactions with boys.

These teachers' implicit bias led them to reinforce the status quo with the girls ("Fine. Good") while propelling the boys to think bigger ("Good. And why don't you try this?").

Implicit biases are those that we carry with us from our upbringing and exposure to our surroundings. They shape our attitudes toward people even without our conscious knowledge. We know a lot about implicit bias now, but in the early nineties, the concept was revolutionary.

I was horrified to learn how implicit bias affected girls from early on. In elementary school, although girls got better grades than boys in science and mathematics, they perceived themselves as doing worse. This perception eats away at their confidence, causing the divide between boys' and girls' interest in STEM that is still prevalent in middle schools and high schools today.

What causes this lack of confidence? There are certainly the not-so-subtle biases, like my high school assuming girls should take home economics and boys should take power tech. But there are

---

49 Myra Sadker and David Sadker, *Failing at Fairness: How America's Schools Cheat Girls* (New York: Charles Scribner's Sons, 1994).

countless small comments and actions that discourage girls from pursuing STEM fields. Now we know them as microaggressions.

My friends and I decided to do something about it.

The Solution: Flexing the Invention Muscle

At MIT, there is a January term. I think it includes "real" courses, but my exposure had been taking a wine-tasting class and sending my younger brother who was visiting to a *Star Wars* marathon. Anyone can propose an activity.

So in January 1993, a few of us proposed a science and technology mentoring program for middle school girls. We called it Keys to Empowering Youth, or KEYs.[50] The program included three components: stereotype awareness, problem solving, and hands-on STEM activities.

Stereotype awareness got the girls questioning the gender gap. In one activity, they cut out magazine pictures that showed how women are portrayed and then talked about the implications for what careers girls choose. For problem solving, we designed activities like my favorite one, which was dubbed Apollo 13 from the scene in the movie where they dump a bunch of stuff on a table that the ground engineers have to use to solve a potentially catastrophic problem in space. For our version, we poured a bunch of stuff on a table and said, "Use this stuff to turn on a light switch from three feet away."

Hands-on activities varied depending on which MIT graduate students were volunteering. The girls might visit a wind tunnel, tour a chemistry lab, or one of my favorites, make a hologram at the MIT media lab.

---

50 "KEYs," MIT Public Service Center, Massachusetts Institute of Technology, accessed March 18, 2022, https://psc25.mit.edu/inspirations/keys.

Our project grew from one independent activities period to a multi-school program. Over a thousand girls participated in the first five years alone.[51] As participating grad students dispersed to other universities, they started new chapters. Today, many thousands have gone through KEYs. And at least until the pandemic, it was still run every year at MIT and the University of Maryland, College Park.

After the report that had inspired all this, Myra Sadker went on to author the first popular book on sexism in 1994: *Failing at Fairness: How America's Schools Cheat Girls.* In the worst of ironies, she died just one year later while undergoing treatment for breast cancer.

The KEYs program showed me firsthand that inviting young girls into the world of STEM could build confidence and change attitudes and that as mentors, we could inspire the girls to pursue STEM careers. But I wanted to do more.

The Power of *Coopertition*

I learned of an opportunity to work with high school kids from a wonderful professor at MIT, Woodie Flowers. Woodie had partnered with inventor Dean Kamen to found a program to make science and technology more inviting to all students. They called it For Inspiration and Recognition of Science and Technology, or FIRST Robotics, and held their first competitive event in 1992.

Dean Kamen is probably one of today's most prolific and impactful inventors, with over a thousand patents to his name. He wants teens to look up to scientists the way they view movie and rock stars.

---

51 Elizabeth Thompson et. al., "Keys to Empowering Youth: Developing A Science and Technology Mentoring Program," *Women in Engineering ProActive Network* (1999). https://journals.psu.edu/wepan/article/view/58146/57834.

As Dean puts it, the goal is "to transform our culture by creating a world where science and technology are celebrated and where young people dream of becoming science and technology leaders."

In fact, the first time I went to a FIRST competition, it felt more like a rock concert than a science fair! A two-robot alliance competed against two other robots in a small arena. Unlike "battle bots," you had to be very careful how you treated the other robots—the alliances were random, and in the next round, you could be paired with your current competition. Dean called this "coopertition" and it was part of the culture of "gracious professionalism," a term coined by Woodie that described the respectful behavior expected of all teams.

When I first started AnthroTronix in 1999, I was a consultant to FIRST, so I got to see the inner workings. We launched the first workshop for girls at FIRST headquarters in New Hampshire. This was around the same time that FIRST partnered with Lego to bring robot-making to even younger kids, down to the third-grade level.

REACHING YOUNGER STUDENTS

When my older daughter reached third grade, I started a FIRST Lego League Jr. team for her and some friends. As a lover of Lego, this was perfect for her! The challenge asked the kids to learn about simple machines like levers or inclined planes and to use them in the design of their solution to a given problem.

One year, the challenge was "Snack Attack—Keeping Food Safe" and took kids into the nitty-gritty of food logistics. They had to develop a presentation and build a model showing where their favorite snack came from and how it got to them safely. They also had to incorporate one moving part and at least one simple machine. Another year, the challenge was right up my alley: "Body Forward—Engineering Meets Medicine." This time, teams had to pick a medical device and build a model showing where and how it works.

REMOVING ECONOMIC BARRIERS

One drawback of the FIRST high school robotics program was that standing up a team was expensive. So another solution was created by Tony Norman and Bob Mimlitch, the founders of Innovation First. They developed a lower-cost robotics toolkit to market and spun off a new competition called VEX Robotics. Both VEX and FIRST have grown worldwide and together engage thousands of kids in robotics competitions each year.

When my daughter hit middle school, we graduated to VEX IQ. VEX IQ uses snap-together parts with greater flexibility but doesn't require a whole garage of tools to build a robot. My daughter and I assembled a team three years running. One year, our team cleaned up at the regional competition, winning both the high point score and the team competition. But mostly, we just had a lot of fun building robots and trying to make them go where we wanted them to—we were exercising that invention muscle.

AnthroTronix not only sponsored and provided engineering mentors for my daughter's VEX team but also worked with a high school FIRST robotics team in DC, coached by my mother! The local news came and did a story on three generations of roboticists.[52]

Calling All Makers

The first Maker Faire was in 2006, and within a few years the maker movement took hold across the country. It's the idea that we are *all* creators and can make stuff. Maker Faires cropped up all over the United States, with the big ones in San Francisco and New York. The movement encapsulated everything I'd been doing with my

---

52 Leon Harris, "Harris' Heroes: Local engineer helps young girls build robots," *ABC7*, February 19th, 2014, https://wjla.com/features/harris-heroes/harris-heroes-engineer-helps-girls-build-robots-100385.

STEM outreach, so when Silver Spring became the site of the first DC-region Maker Faire in 2013, AnthroTronix signed on as a sponsor. That year, 8,000 people of all ages attended; by 2019, attendance was over 15,000.

The organizer of the Maker Faire was Cara Lesser, the founder of a maker space and experiential learning center for children called Kids' Innovation & Discovery Museum, aka KID Museum. If any organization embodied invention as creativity, it was KID. Cara and her team coined the phrase "Mind of a maker" as the foundation of their learning philosophy. Their programs revolve around cultivating STEM skills in the context of real-world, creative problem solving. They understood the critical period of third through eighth grade and focused their energies on that age group. Partnering with the Montgomery County Public Schools, in 2018, they launched the Invent the Future Challenge, which asked teams of middle schoolers to invent something to help save our environment.

My younger daughter, whom I could never convince to do the robotics competitions with us, was now in middle school. I hoped this opportunity would get her to flex that invention muscle, so I offered to coach a team. She was at the public school magnet for performing arts, so it was even more interesting to me to see what these four young makers would come up with.

The competition culminated in a science fair–like expo, with 150 teams from fifty schools around the county. There were projects titled "Recycling" and "Clean Water," with imaginative and clever inventions like a "smart tree" for early detection of forest fires or composting bins that also cleaned water.

I guess my team got a little carried away, as their project sounded like the title of a PhD thesis: "Detection and Mitigation of Aflatoxins in Crops." The girls had read news articles about peanut butter having high levels of these toxins due to mold, which sent us down a path of investigation. Their solution was a drone that could fly over crops at night, shining UV light to prevent molds from developing.

That is the mind of a maker! And yes, they did win one of four "IDEA" category awards. But I think they were more excited about the free doughnuts and Six Flags tickets they won.

Technology competitions are great for showing what humans do well and what machines do well. Like my virtual toast with my friend Jerry at XPRIZE, these competitions also demonstrate the potential for what they can do together. There is an oft-cited competition, Freestyle Chess, which allows teams of any size and makeup, human or AI or some combination. In 2005, a team made up of average chess players with some laptops beat the top human *and* AI players in the world. They had figured out how to leverage computing power to make their whole more than the sum of their parts.

But forget about prizes and winning. It's all about flexing the maker muscle.

MAKERS WITHOUT BORDERS

As I told you in Chapter 8, becoming a member of the Young Global Leaders at the World Economic Forum expanded my view of what I could and should be doing with my career. At the time, I was excited about launching AnthroTronix and bringing Cosmo's Learning System to market but also mindful that this was only a tiny corner of the world that would have access to the technology I was creating. After meeting so many innovators from around the world, I thought, given my expertise and interest, how can I act globally as well as locally?

Another Swarthmore graduate and my longtime friend Dr. Naomi Chesler, now a biomedical engineering professor at the University of California, Irvine, told me about Engineering World Health (EWH). Since 2001, EWH has been sending engineering college students to fix donated medical equipment in several countries in Africa and Central America. That is because 70 percent of the equipment is not in use. Sometimes it is broken and may need a new part. Other times, it is in perfect shape but requires the wrong voltage for that country's power system or the user manual is in the wrong language.

I learned that EWH focused on teaching students to use local resources to service the equipment. For example, blue bili lights are a type of light therapy used to treat newborn babies with jaundice. Jaundice is caused by high levels of bilirubin in the blood, making a baby's skin look yellow, and can cause serious long-term effects if not treated. Replacement bulbs can be difficult to find, but in Nicaragua for example, it's easy to get blue LED strips to use in place of the bulbs because they're often used for motorcycle accent lighting. Another example is to use bicycle inner tube patches for small blood pressure cuff leaks.

I joined the board of EWH and took one of my daughters with me to Nicaragua for a board meeting so she could meet some of the EWH students and see firsthand how invention in other countries may look different than it does in the United States. Often, the students return to their universities and are inspired to enter the EWH design competition, which invites students to submit innovative designs for medical technology that can make a difference. Design teams can choose to work on any project relevant to medical innovations for low-resource healthcare. Sometimes students have ideas based on what they've experienced firsthand during their time in the local hospitals, but EWH also posts the five challenges for medical equipment in low-resource settings so that students can brainstorm new ideas based on documented needs.[53]

Recent winning teams were from the University of Texas at Austin, who designed a low-cost automated way to detect leukemia from blood smear images, and from University College Dublin, Ire-

---

53 "Five Challenges," Engineering World Health, last updated 2022, https://www.ewh.org/five-challenges.

land, who designed a device to deliver emergency epileptic seizure medication through the cheek.

In a wonderful full-circle story, my former student, one of the co-founders of AnthroTronix, Dr. Mike Tracey, joined me on the board of EWH and is now the chairperson. He joins me in supporting these students who, both on paper and in practice, are creating their own maker stories.

Representation Matters

When I founded KEYs, I wasn't trying to turn every girl that came through into an engineer. Nor are any of the organizations I've volunteered with over the years. Instead, they get kids excited about what technology can do. Whether or not they pursue a career in STEM, hopefully the experience makes them see these fields as accessible and creative. Most importantly, they encourage kids to see themselves as makers throughout their lives and to be confident in using technology to solve problems whenever and wherever life takes them.

In 2019, I had the honor of being one of 125 women named an American Association for the Advancement of Science IF/THEN[54] Ambassador. The vision of AAAS was IF we support a woman in STEM, THEN she can change the world. They asked each STEM ambassador to create our own IF/THEN message. Mine was, IF you show girls the power of technology, THEN they can create the future. You've seen more of my IF/THEN messages at the beginning of each book chapter. Now you know what they are!

The originator of the IF/THEN initiative is another larger-than-life woman, Lyda Hill. A successful innovator and entrepreneur, she

54 IF/THEN® is a trademark of Lyda Hill Philanthropies.

was one of the first women to sign The Giving Pledge,[55] "a promise by the world's wealthiest individuals and families to dedicate the majority of their wealth to charitable causes."

A major goal of the IF/THEN initiative was to create a collection of images, activities, and information on contemporary women in STEM, particularly for K–12 education.[56] As the largest database of its kind, no one would ever again be able to say that they couldn't find enough information on women and underrepresented minorities in STEM.

Thanks to this program, we ambassadors are now part of a series of Girl Scout STEM badges, museum exhibits, book content, collecting cards, and TV shows.[57] One example is the Emmy-nominated CBS TV show *Mission Unstoppable*, produced in partnership with the Geena Davis Institute on Gender Studies, which is aimed at getting middle school girls excited about STEM. I was invited to do a segment on inventing and because it was filmed during COVID-19, the one other inventor who was allowed to participate was my daughter![58]

Lyda's vision also included creating #IfThenSheCan—The Exhibit, the largest collection of statues of women ever assembled. This collection of 120 life-size 3D-printed statues of contemporary women in STEM opened Women FUTURES month at the Smithsonian National Mall in Washington, DC in March 2022.

---

55 "The Giving Pledge," last updated 2021, https://givingpledge.org/.
56 "Featured Categories," If/Then Collection, last updated 2022, https://www.ifthencollection.org.
57 "IF/THEN® Contemporary Women in STEM," International Museum of Surgical Science, accessed March 18, 2022, https://imss.org/if-then-contemporary-women-in-stem/.
58 Mission Unstoppable, "Inventing the Future with Dr. Cori Lathan | Mission Unstoppable," YouTube video, 4:47, March 1, 2021, https://www.youtube.com/watch?v=zXvRlwsU7Ro.

A life-size statue of me with one of my inventions, String. #IfThenSheCan—The Exhibit. (Photos courtesy of Julia Borum.)

In addition to being honored to be included, I love being surrounded by so many kickass women in STEM. Being part of a posse of women statues is a far cry from being the only woman in so many other situations throughout my life. And even though it's only a snapshot in time, hopefully, like technology, it's a bridge to what's to come in the future.

Inspiration Goes Both Ways

I'm inspired every day by my colleagues, friends, former students, and young makers out there. The creativity and enthusiasm of the new generation of makers fuels my desire to keep pushing the boundaries of what technology can do to create a better tomorrow.

My mother inspired me on my professional journey, and I'd like to believe that I inspired her as well. After all, she delved into robotics because of me! I see the same thing happening with my teenagers. I'd like to believe that I inspire them even if sometimes they won't admit it. They certainly teach me and inspire me—for example, with their commitments to racial justice and LGBTQ+ inclusion. I asked them to come up with their own IF/THEN quotes, and here's what they said:

IF you put yourself in someone else's shoes, THEN we might achieve true justice.

—Lindsey, 18

IF you treat people like they matter, THEN others will do the same.

—Eliza, 16

My daughters also helped channel their grandmother so that we could give my mother the last word:

IF you recognize the beauty in $e^{\pi i}$=-1, THEN you understand the profound nature of the Universe.

My mother would have said "the profound nature of God," but she's not here to correct me, so I guess I get the last word after all!

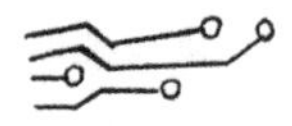

I hope this inspires you to invent your own IF/THEN. Post it on social media and tag #IfThenSheCan, or email it to me at cori@inventthefuture.tech.

IMAGINE
INVENT
BEGIN

CONCLUSION

Joy of the Journey

INVENTING IS LIKE writing a play, or composing a musical piece, or painting a piece of artwork. The invention may or may not turn into a product, and the same is true for works of music, art, and theater.

First, you have this amazing idea—a story, a melody, or a way to solve a problem.

Then it manifests. The creative process takes hold, you develop characters, and the plot unfolds, bringing the characters to life. With technology, those characters might be components you have to find or design. You bring them together in a proof of concept and see what happens when they interact.

And that may be where it ends.

And that's awesome!

In the invention world, we call this a bench prototype. There are many artists who create their art for themselves or create many iterations of art before they turn outward. My roommate from college writes screenplays, and she says she envisions JesterBot in a box with some of her favorite screenplays that have been read by only a few close friends.

Or maybe your screenplay is ready for actors and an audience. That takes it to a whole new emotional and physical level, transforming the story to screen and playing with elements like production design, music, and editing. With an invention, the people who can benefit from it get to try it out. With both a screenplay and an invention, you get to see *impact,* and that is an amazing feeling.

And that may be where it ends.

And that's awesome!

The name for this stage of invention is a field or beta prototype. The VISUnit and other inventions at AnthroTronix got to this point, expanding our understanding of what was possible and enabling us to drive the technology forward.

Then, depending on the invention—or the screenplay—you might be able to get the funding to distribute it. Maybe it's the suspenseful drama you envisioned from the start, but audiences are in the mood for a comedy that season. Or maybe it becomes a cult hit in the streaming market. When we started with DANA and the military, we knew the device was important, but we never would have predicted the path would lead to civilian healthcare companies.

And that's awesome too!

Most inventions may inspire only a few people, such as other inventors, who then go on to create and inspire others. To me, that's the most powerful reason to invent. The very act promotes shared experiences, and I believe those connections are the key to improving the world. Whether through collaboration, consumption, or reflection, our shared experiences show us what's possible and how our actions affect each other.

What drives me to create? I suspect it comes down to my passion for manifesting something tangible out of something intangible. For harnessing the power of technology to increase our real abilities as humans. And for taking us closer to human–machine collaborations that bring out the best in both.

What drives you? I truly believe that no matter where your skills and interests lie, practicing the art of invention will expand your conception of who you are, what is possible, and why technology matters.

Take the risk. Make the thing. Find cool people to work with. I can't guarantee where your next invention will lead, but I promise you this:

IF you harness technology to improve lives,

THEN you will create a future you love.

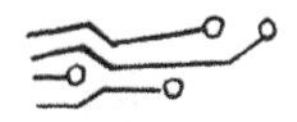

ACKNOWLEDGMENTS

This book is first and foremost a "legacy memoir" to tell the story of AnthroTronix after over twenty years of invention. Of course, it is my story as well and the story of so many of our employees, interns, collaborators, and professional and personal colleagues. Unfortunately, I had to leave out twice as many stories as went in since a book can only be so long. But maybe there will be an *Inventing the Future Part II*!

In any case, I want to use this space to focus on the people who were so critical to AnthroTronix' success over the years.

That said, I'll start by acknowledging and thanking the most important non-AnthroTronix person, my husband and soulmate, David Kubalak, who supported me in my adventure to start a company and then the crazy journey that followed. Dave, I think you would say, "How is that different than being with you on any of our other crazy adventures!" Not the least of which are raising two amazing children, Lindsey and Eliza, who light up my world with their smiles even when they are simultaneously driving me to distraction.

I also want to recognize my parents who are reflected throughout the book as they were my biggest supporters until they were no more. On my mother's literal deathbed in the hospital, when she couldn't even speak, she still scrawled notes as I read her my latest chapter draft.

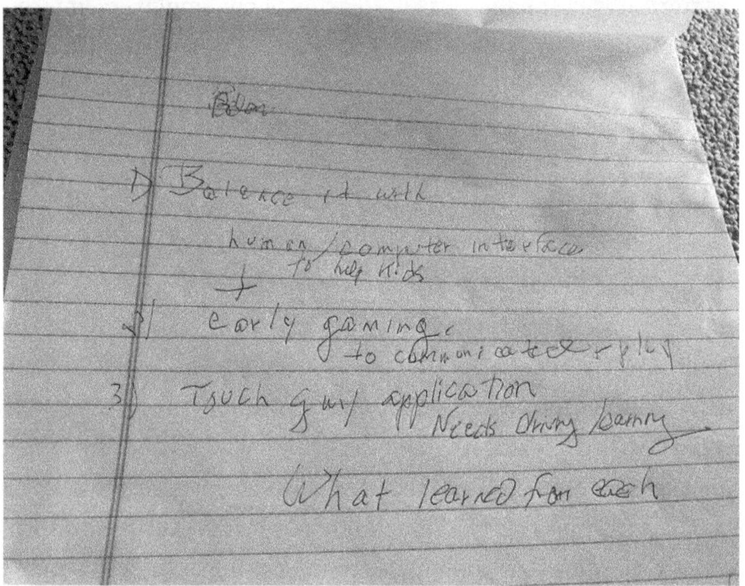

Thank you to Yo-Yo Ma for being a friend and agreeing to share your wisdom through writing the Foreword to this book. Thank you for choosing optimism and for using the power of culture to make positive change.

Thank you to the Scribe Tribe, whom I saw almost every week for two years. Thank you to those who went before me—you inspired me to keep going. And to those who are coming along behind me—you can do it! To all the amazing folks at Scribe, especially Emily G., Chas H., Meghan M., and Joy Y.—Thank you for your advice and simultaneous ability to whip-crack *and* calm me. To Hussein A., I *may* have been able to do this without you, but I wouldn't have wanted to! Thanks for being a mentor and a friend. (Please read Hussein's amazing story as a refugee, *Art of Resilience*, and support at refutees.com.)

To my "editorial team," Grace Bulger and MT Cozzola, thank you both for holding my hand throughout the process. Grace, thank you for the many late-night conversations working through a story or a theme. MT, working with you collaboratively made this book better and more fun to write. Thanks to Jane Metcalf for your help with web design and book launch.

I don't remember how we found Kate Rutter many years ago to work on our AnthroTronix graphics, but we finally got to meet and work together on the graphics for this book. Kate, thank you for putting your brilliance and talent into the book and for manifesting my vision for each chapter in your art.

The book cover image was based on the "DivaVinci" Vitruvian Woman that launched with AnthroTronix many years ago. Thank you Kate and Peter for bringing her into the future and to Rebecca Lown for being the most patient cover designer on earth!

I had so many wonderful volunteers to read parts and sometimes all of the book. To my alpha and beta readers, thank you! Annie B., Kanali B., Samantha B., Bridie H., Rebecca H., Laamia I., Megan L., Lara M., Devyani M., Daniela M., Kelly O., Sarah P., Vishal R., Bo R., Jason S., Jeannie S., Brendan T., Eli and Stephanie V., Guiliana W., Jessica X.

I also want to give a shout-out to Arthur Daemmrich and Joyce Bedi of the Lemelson Center on Innovation and extraordinary jour-

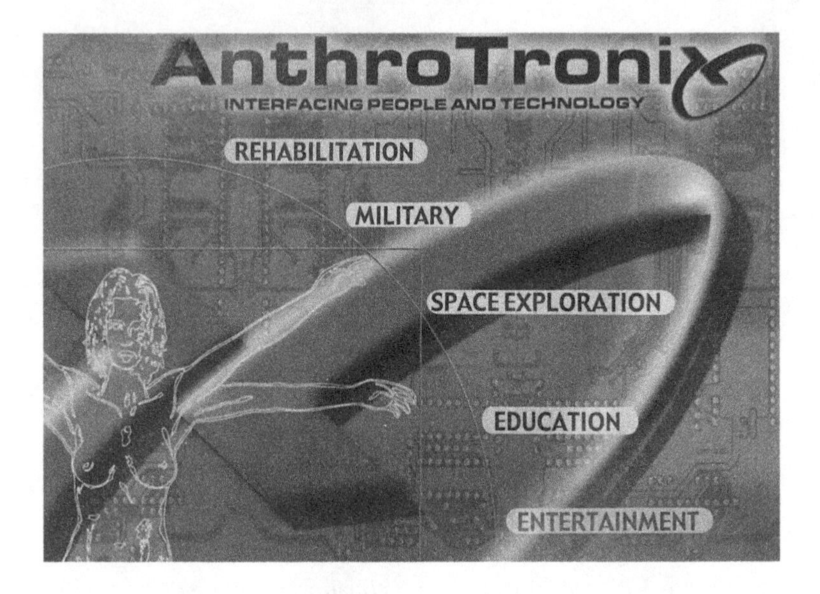

nalists Joe Palca and Mariette DiChristina, who all gave me feedback and great advice to help set the tone and positioning of the book. Thank you, Dr. Andrew Maynard, for your continuing feedback and for inspiring me with your books *Films from the Future* and *Future is Rising*. Thank you to Cindi Leive for your advice and wisdom—you are always a source of inspiration, including your latest venture to fight for gender and racial justice and equity at www.WeAreTheMeteor.com!

Now to the bulk of my acknowledgments—my AnthroTronix family and extended colleagues!

Starting AnthroTronix was a leap of faith, and I couldn't have done it without Mike Tracey and Jack Vice. Mike and Jack, we had some crazy times early on as a maverick startup, from MMVR conferences and rock 'n' roll science with kids, to Quantico and real-time operational exercises with soldiers. Thank you both for being my partners in crime to launch AnthroTronix. I'm honored to still call you both colleagues and friends! Not always true with co-founders.

To my mentor, advisor, and voice of reason Carl Pompei. You brought experience, structure, and a lot of patience to AnthroTronix early on. Thanks for getting that chip off my shoulder after so many failed "advisors." And thank you to you and Penny for being the foundation of our AnthroTronix family.

And if Carl is the foundation, Charlotte Safos must be our glue! Charlotte, you have been my right-hand person for so long, I can barely remember AnthroTronix without you. But I do remember meeting you at Lauren's wedding, and the rest is history!

Jen Story—what metaphor can I use? The engine? Thank you for keeping us running, which is no small feat.

Marissa Lee—in addition to all you do as a critical part of the DANA team, it was amazing to work with you on the book. Knowing you had my back on all the little but important stuff kept me focused and working toward the goal. Thank you also for shaping, researching, and even writing drafts of reflections, which moved them forward so much faster than they would have otherwise.

James Drane—you have incredible discipline and rigor as an engineer and program manager *and* a commitment to work-life balance. Respect. Tiffany Worthy and Kevin Haines—Thanks for being part of the DANA team and sticking with AnthroTronix through the pandemic. There are good things to come!

Many of our employees over the past twenty years stayed for over a decade. Thank you, Regina Fraiser, for coming onboard "just to help me get set up"; thank you, Justin Rightmyer, for giving AnthroTronix your excellent mind and brave spirit; thank you, Amy Brisben, who for years embodied the heart of CosmoBot and Cosmo's Learning System. Thank you, Jon Farris, for helping transform brain health through launching DANA among other things.

Thank you to Rita Shewbridge who for years managed our intern program—over a 100 students from MIT, Swarthmore, UMD, NASA Goddard's summer academy, local high schools, and more. Thank you to our AnthroTronix longtime colleague Sanjay Mishra who, among other things, brought our vision for 3D Space alive by being so much more than a computer programmer.

Dr. Anna Skinner—I can't believe how far back we go and how many stories we could tell, including being saved by an angel. The one paragraph I was able to include in Chapter 8 on INSPIRE STEM doesn't even scratch the surface of what you built at AnthroTronix and now for yourself as CEO of Black Moon Consulting. I will always laugh thinking about your brilliant acronyms for project names, which *always* got funded—DROID, SUSTAIN, INTERACT, INSTRUCT, MUSTER, etc. Thank you for helping me recall the stories even if many of them didn't make it into the book. Let's have a drink and remember some more!

Many of our employees started or built their careers at AnthroTronix before moving on to their next big adventure, which we always celebrated! Thank you for being part of our journey: Lisa B., Jeff B., Ian C., Kris E., Patrick G., Lily G., Ara H., Bryan H., Hiroshi I., Ben L., Tim M., Josh N., Matt P., Emma R., Kyle R., Brandon R., Clementina R., Alem S., and Lawrence W.

I also want to thank the many AnthroTronix advisors we had over the years: Dr. Murali Doraiswamy and Dr. Adam Kaplin for your science and clinical advice; three successful business women—Amy Butte, Ruma Bose, and Linda Avey—for your friendship and mentorship! To Matt Patterson and Steve Sidel, thanks for trying to get me to focus when I just wanted to invent more stuff. And special thanks to our advisor Pam Melroy, who became my fast friend over brunch in Houston many years ago and then joined the AnthroTronix board and helped us to navigate difficult times.

I also want to thank our initial angel investors who believed in the potential for CosmoBot and Cosmo's Learning Systems over twenty years ago when we first launched the company—Kit G., Jeff H., Brian H., Ed L., Lissa M., Daniel M., Pam M., Bob M., Tony N., Ray S., Dave W., Chris W., Todd and Chris W., Kristin Z.—I hope we've made you proud with the number of people we've helped and inspired even if we didn't make you any money. :)

Each chapter could have been filled with the names of our collaborators and related projects, but that obviously would have been distracting. So in the interest of good storytelling, I had to simplify. What follows is my attempt to do justice to some of the folks and projects that I didn't mention above. To those whom I've already included in the stories, thank you for contributing your time and ideas to my book. I am indebted to your work to invent a great future and this book wouldn't be the same without you!

In Chapters 1 and 2, you met a few of the people critical to CosmoBot and Cosmo's Learning System, but just like I didn't build ATinc alone, they had colleagues who were part of the story as well. Along with Dave Warner, Syracuse physics professor, Dr. Ed Lipson was co-director of the Center for Really Neat Research and co-founder of MindTel, Inc., one of our first investors. Thanks, Ed and Dave, for your support! We worked closely with the University of Maryland's Human Computer Interface Laboratory Director Dr. Allison Druin and Associate Director Dr. Catherine Plaisant. Thank you, Allison and Catherine, for your work including the voices of kids in design. We loved working with you!

Our partners at the Mayo Clinic's Motion Analysis Lab included Dr. Krista Coleman and lab director Dr. Kenton Kaufman. Thank you both for being open to innovation!

There are so many untold stories from working with the amazing National Rehabilitation Hospital. Thank you to former directors of the Assistive Technology Laboratory, Bill Peterson and Dr. Mike Rosen, for recognizing and building on the potential for virtual reality, sensors, robotics, and more, before it became a thing!

Working with kids with autism, we had amazing colleagues in the late great Dr. Cheryl Trepagnier; Dr. Katharina Boser who is still a maker extraordinaire; and Dr. Carole Samango-Sprouse, Director

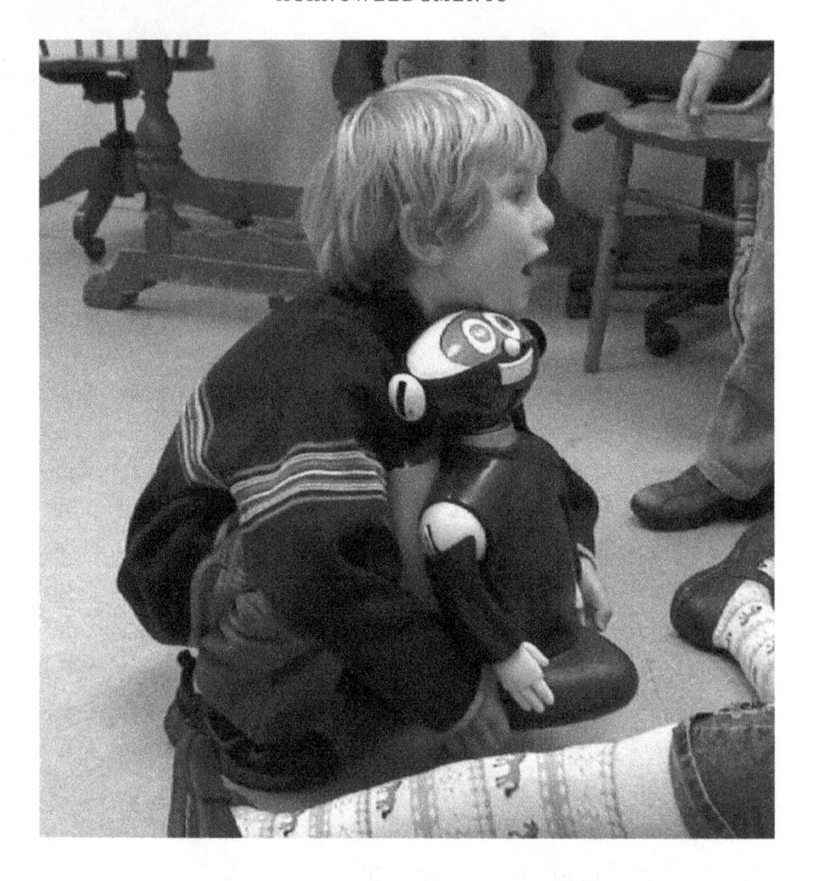

of Neurodevelopmental Diagnostic Center for Children, which was the site of so much of our work with kids.

Cosmo's Learning System and AT KidSystems had great leadership in Judd Nathanson and Erin Lavin, who brought their expertise in manufacturing and education, respectively, to the table. And a big thanks to the dedicated therapists who worked with so many kids on reaching their full potential, especially Jen Bilyew and Elaine Schwartz.

And of course, a huge thanks to the kids themselves who are mostly all adults now and I hope have great memories of working with both real and virtual CosmoBot!

Chapter 4 and the following interlude were jam-packed with amazing folks I had the opportunity to work with, but there were some key folks I could only allude to despite their important roles in the stories. In particular, I want to thank my first real mentors Drs. Paul DiZio and Jim Lackner. Jim and Paul are Co-Directors of the Ashton Graybiel Lab at Brandeis University, where I fell in love with scientific research. They set a high bar for those who followed! Dr. Gilles Clément was one of those mentors, as well as (the late) Dr. Larry Young and Dr. Chuck Oman at the Human Systems Laboratory Lab at MIT and my BCS MIT mentors Dr. Emilio Bizzi, Dr. Jenny McFarland, and Dr. Conrad Wall. Thank you all for nurturing my passion for science and engineering!

Finally, thank you to my friend and mentor Dr. Dava Newman for being an incredible leader and role model for so many of us.

Chapter 5 centered on the VisUnit story, which wouldn't have happened without our partners at Lockheed-Martin Advanced Technology Laboratories. Our point person was Brett Breslow, whom we met during AugCog (Chapter 8) as our star subject given he had no hair and could easily wear an EEG cap. :) Thank you, Brett, for working with us to achieve the vision to advance technology for our troops albeit many years ahead of its time. And speaking of AugCog, I simplified a few years of crazy DARPA program activity into one paragraph, so I'll do the shout-outs and acknowledgments here. Thank you to Dr. Dylan Schmorrow, the DARPA PM who conceived of AugCog and launched a field! AugCog is also where we met Dr. Amy Kruse. Thank you, Amy, for your vision(s!) then and now. Thank you to the project PI Gerry Mayer for giving us the opportunity to work with your amazing team including a trio of kickass women—Dr. Polly Tremoulet, Dr. Susan Regli, and Dr. Jodi Daniels. Jodi is now Commanding General of the US Army Reserve Command, the first woman to serve in that role!

I also want to thank my colleagues at PTC, especially CTO Steve Dertien, whom I can always count on to help me think about how to communicate a future where the digital and physical converge.

The DANA project described in Chapter 6 was such a large undertaking that I will never be able to thank everyone individually. We had incredible military partners whom we worked with to develop DANA in operational environments. Captain Jack Tsao had the vision, and many helped take it forward into Afghanistan, shipboard, and implement it with convoy drivers and EOD training, among other things. Our research and early commercial adopters of DANA are numerous, but we never would have transitioned to the clinic if it hadn't been for Dr. Adam Kaplin's belief in the clinical power of DANA.

I want to recognize and thank early telemedicine pioneer Stephen Wyle, who spun out a company to commercialize the Telemedicine Instrumentation Pack we used during Strong Angel. TIP, like so many of the inventions I talked about, was way ahead of its time but demonstrated a bright future.

Chapter 7: Thanks to Dr. Kevin Tracey and the team at the Feinstein Institutes for Research at Northwell for your past and present leadership excellence in clinical research. To Dr. Martine Rothblatt, Paul Mahon, and the United Therapeutics team, thank you for believing in the vision of electroceuticals and moving the vision forward. And to Dr. Kate Rosenbluth and Renee Ryan and their Cala Labs team for including us on the journey to bring the dream of electroceuticals to fruition. I look forward to what we can do together!

My description of the World Economic Forum in Chapter 8 and the impact of being part of the Young Global Leader community couldn't do justice to describing the impact that being part of that global family has had on me. Everywhere I look, my life is enriched by YGLs. Case in point were my brilliant panelists for the Kennedy Center Arts and Science Summit that I opened the chapter with—neuroscientist Dr. Olivier Oullier and artist Drue Kataoka.

It's impossible to acknowledge everyone individually, so in addition to a heartfelt thanks to the entire community, a special shout-out to the DC YGLs (past and present!) and the Costa Rica crew—you know who you are. Finally, thank you, April Rinne, for your book *Flux: 8 Superpowers for Thriving in Constant Change,* not only for the timely content but also for sharing your personal story of loss in the book and with me personally.

Thank you to The Weekend tribe—I may answer to "robot," but I'm still a villager! And a special shout-out to Weekend founder, YGL, and friend Nancy Lublin—Can I join you in your vision, for our next fifty years, to help each other?

Chapter 9 was filled with those I need to thank who are promoting the mind of a maker. To the teams at XPRIZE, FIRST, and the Robotics Education and Competition Foundation, thank you for providing venues for us all to exercise our maker muscle! Thank you to Engineering World Health for helping to invent a future of equitable access to medical technology. Thanks again to Dr. Naomi Chesler for introducing me to EWH and for your own work at BuildingStemEquity.com in increasing STEM creativity through best practices in diversity, equity, and inclusion.

To the Society of Women Engineers for supporting the KEYs program and UMD and MIT for carrying the torch.

Thank you to the great folks at AAAS and Lyda Hill Philanthropies and Geena Davis Media Institute for their support and partnership on the IF/THEN initiative. The positive If/Then statements were exactly the optimistic framework I needed for my book. Thank you, Lyda Hill, for fulfilling a bucket list dream I didn't even know I had and making me into a statue! Seeing kids (and adults) recognize me next to my statue in the Smithsonian's Futures Exhibit and then talking to me about my work was indescribable. And not only was I surrounded by 120 other women statues in STEM but also by a tribe of women friends who came to take selfies with me and/or my statue including Ariana, Cara, Elaine, Gina, Grace, Heli, Julia,

Kathleen, Nilmini, Rachel, Sarah, Suzy, and Tina. Finally, thank you, Rachel Goslins, for making the Futures Exhibit come alive and including me in that journey, and Dr. Ellen Stofan, for your ongoing leadership in STEM, and to you both for being such great friends!

To the If/Then Ambassadors—I celebrate and thank each of you for all you do as women in STEM. And that journey comes with the good and the bad, so let's continue to surround each other as our statues did and to hold each other up. Although I can't list you all, I wanted to call out some of you who were able to support my If/Then "Maker Girl" virtual workshops or were part of the writer's group—Adrienne, Afua, Bea, Charita, Danielle, Davina, Heather, Jenny, Karina, Kimberly, Kristen, Mitu, Nicole, Paula, Ronda, Sam, Sarah, and Wendy.

Finally, last but not least, I want to thank Cara Lesser, the founder of the KID Museum; her team at KID; and the amazing KID board—Alex, Antonio, Brian, Charles, Chris, David, Jill, Josh, Michael, Robby, Sally, Sam, Shanika, and Sue. Thank you all for helping thousands of kids exercise their maker muscle and so invent the future every day!

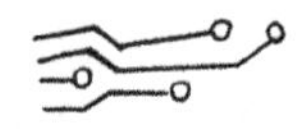

ABOUT THE AUTHOR

Dr. Corinna (Cori) Lathan is a technology entrepreneur who has invented robots for kids with disabilities, virtual reality technology for the space station, and wearable sensors for training surgeons and soldiers. She co-founded the biomedical engineering company AnthroTronix, Inc. and for over twenty years has led it as CEO and Board Chair.

She received her BA in biopsychology and mathematics from Swarthmore College and an M.S. in Aeronautics and Astronautics and Ph.D in Neuroscience from the Massachusetts Institute of Technology.

Cori is a passionate advocate for STEM (science, technology, engineering, and mathematics) education, global health, and LGBTQ+ inclusion. She lives in Silver Spring, Maryland with her husband Dave, two children - Lindsey and Eliza, two dogs, three cats, and a varying number of fish. Their family motto is, "We may be crazy, but we're not stupid."

Depending on her mood, she may be reached in one of the following ways:

cori@inventthefuture.tech
linkedin/in/clathan
@clathan (Twitter)
@drcoril (Instagram)